Lead Them with Virtue

Lead Them with Virtue

A Confucian Alternative to War

Kurtis Hagen

LEXINGTON BOOKS
Lanham • Boulder • New York • London

Published by Lexington Books
An imprint of The Rowman & Littlefield Publishing Group, Inc.
4501 Forbes Boulevard, Suite 200, Lanham, Maryland 20706
www.rowman.com

6 Tinworth Street, London SE11 5AL, United Kingdom

Copyright © 2021 The Rowman & Littlefield Publishing Group, Inc

All rights reserved. No part of this book may be reproduced in any form or by any electronic or mechanical means, including information storage and retrieval systems, without written permission from the publisher, except by a reviewer who may quote passages in a review.

British Library Cataloguing in Publication Information Available

Library of Congress Cataloging-in-Publication Data

Names: Hagen, Kurtis, author.
Title: Lead them with virtue : a Confucian alternative to war / Kurtis Hagen.
Other titles: Confucian alternative to war
Description: Lanham : Lexington Books, [2021] | Includes bibliographical references. | Summary: "Kurtis Hagen argues that early Confucians seek to discourage war by prescribing conditions for just war that are exceedingly difficult to meet. They encourage, instead, a long-term strategy of ameliorating unjust circumstances by leveraging the credibility and influence that stems from consistently practicing genuinely benevolent governance"— Provided by publisher.
Identifiers: LCCN 2021024244 (print) | LCCN 2021024245 (ebook) | ISBN 9781793639707 (cloth) | ISBN 9781793639721 (paperback) | ISBN 9781793639714 (ebook)
Subjects: LCSH: War (Philosophy)—History—To 1500. | Military ethics—China—History—To 1500. | Confucianism and state. | Intervention (International law)—Moral and ethical aspects. | Confucian ethics. | Philosophy, Chinese—221 B.C.-960 A.D.
Classification: LCC U21.2 .H33 2021 (print) | LCC U21.2 (ebook) | DDC 355.02—dc23
LC record available at https://lccn.loc.gov/2021024244
LC ebook record available at https://lccn.loc.gov/2021024245

Contents

Preface		vii
Introduction: Confucianism and Noncoercive Moral Leadership		ix
1	A Brief Overview of Confucianism	1
2	Western and Chinese Attitudes Regarding Warfare	11
3	Anticipating Confucian Just War Theory	33
4	Mencius on War and Humanitarian Intervention	55
5	Xunzi on War and Humanitarian Intervention	79
6	Mencius and Xunzi on Tyranny and Humanitarian Intervention: A Response to Twiss and Chan	91
7	From Human Nature to the Clash of Civilizations	113
8	Two Visions of Confucian World Order	131
Conclusion		149
Bibliography		153
Index		161

Preface

This book contains a mixture of new material and revised previously published material.

Previously Published Material

- Chapter 2 is a revised and expanded version of my article "A Chinese Critique on Western Ways of Warfare," *Asian Philosophy* 6.3 (November 1996): 207–217, reprinted by permission of the publisher (Taylor & Francis Ltd, http://www.tandfonline.com).
- Much of the material in chapters 4 and 5 comes from my article "Would Early Confucians Really Support Humanitarian Interventions?" *Philosophy East and West* 66.3 (July 2016): 818–841. University of Hawai'i Press.
- An earlier version of chapter 8 was published as "Project for a New Confucian Century." *A Future Without Borders: Theories and Practices of Cosmopolitan Peacebuilding*, edited by Eddy Souffrant. New York: Brill/Rodopi, 2016, 168–189. (A few paragraphs have been incorporated into chapter 7 rather than chapter 8.)
- Quotations from four chapters that appeared for the first time in the following book are included with permission. From: *Chinese Just War Ethics: Origin, Development, and Dissent*, edited by Ping-cheung Lo and Sumner B. Twiss, copyright © 2015 by Routledge. Reproduced by permission of Taylor & Francis Group. The relevant chapters are: "Varieties of Statecraft and Warfare Ethics in Early China: An Overview" and "Legalism and Offensive Realism in the Chinese Court Debate on Defending National Security 81 BCE," both by Ping-cheung Lo; "Zheng (征) and Zheng (正)? A Daoist Challenge to Punitive Expeditions," by Ellen Y. Zhang; and "Mohist Arguments on War," by Hui-chieh Loy.

A Note on Translations and Conventions

Unless otherwise indicated, translations of the Confucian and Daoist classics are my own. For the military texts, the *Sunzi* and the *Sun Bin*, I use the translation of Ames 1993 (*Sun-Tzu The Art of Warfare*) and Lau and Ames 1996 (*Sun Pin: The Art of Warfare*), respectively.

Quotations from the *Xunzi* include the chapter/page/line number in Lau and Chen's (1996) concordance. The relevant section numbers in John Knoblock's three-volume translation are also included (Knoblock 1988, 1990, 1994). Knoblock's numbering system is also used in *Philosophers of the Warring States: A Sourcebook in Chinese Philosophy* (Hagen and Coutinho 2018). No specific editions are given for the *Mencius* or the *Analects* since relatively standard passage numbering makes it easy to locate the cited passages in most editions.

In quotations, Chinese words that have been Romanized according to systems other than pinyin have been converted to pinyin.

Following relatively common practice, I use the Latinized name "Confucius," for Kongzi (Master Kong), and likewise "Mencius" for Mengzi (Master Meng). All other names are based on the pinyin Romanization of the Chinese name, such as Xunzi. Though convention may be shifting toward the quite reasonable practice of separating the parts of the names, such as Xun Zi (Master Xun) rather than Xunzi, I have herein continued what has been the conventional practice of combining the parts as if a single name. There are other conventions for Romanizing names, which in some cases differ considerably from pinyin. For our purposes it may suffice to be aware that "Hsün Tzu" is another way of writing "Xunzi." (In pinyin, "x" is pronounced as a *sh*.)

Acknowledgments

Portions of this project were supported by a one-year sabbatical leave from SUNY Plattsburgh, 2013–2014. Parts of the book benefited from feedback from anonymous peer reviewers along the way, as well as from David Kratz Mathies and Tongdong Bai. Kent Simmons, a friend and retired philosophy professor, helpfully provided proofreading and general feedback, as did my parents, George and Joann Hagen. The original version of chapter 2 benefited from feedback from Roger Ames, who, as my teacher for many years, also greatly influenced my understanding of Chinese philosophy generally. Those familiar with Ames's contributions to the field may discern his influence in chapter 1 especially. Finally, Meredith Cargill provided detailed critical commentary on several chapters, as well as countless specific suggestions for rhetorical improvements and general editing, for which I am very grateful. Remaining faults are, of course, my own.

Introduction
Confucianism and Noncoercive Moral Leadership

The title of this book, *Lead Them with Virtue* (*dao zhi yi de* 道之以德), comes from *Analects* 2.3, in which Confucius suggests that when prominent people exhibit exemplary conduct and admirably observe ritualized norms of propriety (*li* 禮), others develop a sense of shame and strive to improve themselves. Confucius held that inspiring moral example, aided by guiding norms of propriety, facilitates the achievement of social harmony. He says so repeatedly, for example:

"Governing is being proper. If you lead by being proper, who would dare not be?" (*Analects* 12.17)

"If one is proper in one's own character, what problem will one have governing? If one is *not* able to be proper in one's own character, how can one make other people proper?" (*Analects* 13.13)

"If you are after governance . . . Simply desire excellence yourself, and the common people will be excellent also. An exemplary person's character (*de* 德) is the wind; a petty person's character is the grass. When the wind flows over grass, it is sure to bend." (*Analects* 12.19)

"If one is able to govern a state by means of ritual propriety and deference, what difficulties would there be? If one is unable to govern a state by these means, what purpose would ritual propriety serve?" (*Analects* 4.13)

"If one comports oneself properly, the people will work well, without even being commanded. But if one does not comport oneself properly, even commands will not be followed." (*Analects* 13.6)

"When those above are fond of ritual propriety (*li* 禮), the common people are easily governed." (*Analects* 14.41)

"Guide (*dao* 道) a small state by living up to one's words while respectfully managing affairs, by caring for others while being moderate in [one's own] expenses, and by utilizing the common folk as seasonally appropriate." (*Analects* 1.5)

It is clear that Confucius (and other Confucians as well)[1] thought that prominent people should lead by means of noncoercive moral example. This book argues that Confucius's most influential early followers, Mencius and Xunzi, who spoke more directly about warfare than did Confucius, would support the strategy of leading with virtue as an alternative to military interventions. It further suggests that this is a reasonable position worthy of serious consideration.

CHAPTER BY CHAPTER OVERVIEW

The opening chapter, "A Brief Overview of Confucianism," introduces the three principal early Confucian philosophers, Confucius, Mencius, and Xunzi, and also seven important Confucian concepts (*junzi, de, ren, li, yi, tian*, and *dao*). This will facilitate understanding of the following chapters.

The next chapter, "Western and Chinese Attitudes Regarding Warfare," describes two pervasive Western attitudes regarding war. One is a romanticization of violent conflict, involving the idea of war as a duel on a grand scale. The other is a justification of violence in the name of some transcendent principle: war is Good against Evil. These Western attitudes are then contrasted with a traditional Chinese attitude toward warfare, drawn from Daoist and Mohist sources and the classic Chinese texts on military strategy as well as Confucian texts. The chapter explores potential problems with renouncing the Western sensibilities regarding war and also with maintaining them. Finally, it proposes a "Confucian solution" for maintaining peace and security. This solution is elaborated over the course of the succeeding four chapters, especially chapters 4 through 6.

Chapter 3, "Anticipating Confucian Just War Theory," lays the groundwork for the next three chapters, which focus on Confucian requirements for the justification for war and humanitarian intervention. First, chapter 3 notes a similarity between Mohists and early Confucians—both were *activists*, whose words were intended not merely to say something but to *do* something. If one focuses too much on their content, without due attention to their intent, the application of their ideas may diverge from what they would have wanted.

It is important to remember that, in talking about war, early Confucians, like their Mohist counterparts, were not so much trying to analyze it as to prevent it. Second, if early Confucians are viewed as activists, like their Mohist counterparts, who are trying to avert war, their optimistic rhetoric regarding the abilities of a true king need not be understood as naïve. Third, early Confucians probably understood and shared some concerns expressed in Daoist critiques of war. Insofar as these critiques are reasonable, skepticism about the efficacy of war, even if morally motivated, may be warranted, and alternative approaches may not be as naïve as they might otherwise seem. It may be that, rather, exaggerated rhetoric was adopted in an effort to encourage wiser as well as more humane actions and policies. Forth, continuing to address the purported naïveté of Confucian idealism, this chapter discusses the potential tension between the Confucian imperative to act altruistically so as to achieve desirable results and the imperative to act within certain bounds of propriety in doing so. The question is, "Would Confucians condone violence in service of good consequences?" Here I explain why the answer is "no."

These four considerations set the stage for an examination of Confucian perspectives on humanitarian military intervention, taken up in chapters 4 through 6. There I argue that early Confucians (Mencius and Xunzi in particular) would, and contemporary Confucians *should*, stick to their pacifistic principles, and resist invitations, in the service of "humanitarian" goals, to make exceptions to norms against violence. While I have elsewhere argued that Confucianism offers a flexible ethics,[2] for there are often many appropriate ways to act positively, some *restrictions* on behavior are less flexible. At the end of this chapter, I acknowledge that it is unclear that Confucius himself would endorse the positions I attribute to Mencius and Xunzi. He *may* have been *too* flexible regarding the use of force. But much remains unclear about this.

Chapter 4, *"Mencius on War and Humanitarian Intervention,"* drills down into the Confucian text the *Mencius* and responds to a number of recent "just war" interpretations that suggest that Mencius would support contemporary humanitarian interventions based partly on his supposed support for "punitive expeditions." I argue that it is misleading to suggest that Mencius *encouraged* punitive expeditions. For he indicated approval only of idealized, semi-mythic examples, and generally those examples served to *contrast* with an existing situation with the import being that the ruler should concentrate on governing his own people more humanely. Far from encouraging the use of the military to solve problems, Mencius intervened with rulers who had too much of a tendency to use the military and encouraged them to adopt a different strategy: win over the world through moral governance at home. Ultimately, this chapter argues that Mencius would not support war, in plausible circumstances, even if motivated by humanitarian concerns. If articulated in a just war theory framework, the Mencian version of the theory would be so strict that war would be excluded in all realistic scenarios.

Chapter 5, "Xunzi on War and Humanitarian Intervention," suggests that Xunzi holds a similar view to Mencius, on my interpretation of him. While a *prima facie* case that Xunzi would support humanitarian intervention can easily be made, as an essay by Yi-Ming Yu exemplifies, a more careful treatment of the text tells a different story. I argue that there is an early Confucian consensus—between Mencius and Xunzi, at least—that is pacifistic in its practical implications: military intervention is only justified if it does not involve genuine warfare. And that is only possible if the authority responsible for the intervention has exceptional moral credibility. Except in semi-mythic tales, this kind of exceptional character doesn't exist. In this context it makes sense that, instead of addressing problems with force, Xunzi advocates turning to civil leadership as a long-term strategy for improving conditions, first at home, and ultimately more broadly.

In chapter 6, "Mencius and Xunzi on Tyranny and Humanitarian Intervention," I address the arguments of Sumner Twiss and Jonathan Chan, who treat Mencius and Xunzi together, arguing that they maintained that foreign tyranny or serious misrule would justify, or even require, military intervention. Indeed, according to Twiss and Chan, Mencius and Xunzi were both "especially concerned with the undertaking of punitive expeditions" to address tyranny and aggression (2015b, p. 123). I argue that this is misleading, and that my analysis, provided in the preceding two chapters, makes better sense of Twiss and Chan's own examples when considered carefully.

Arguments made to justify war, and violence more generally, are often disingenuous, and supported by false empirical claims, in the service of ulterior motives. Chapter 7, "From Human Nature to the Clash of Civilizations," explores Confucian considerations related to this reality. It begins with Xunzi's theory of human nature, namely, that natural human dispositions are "detestable," that is, people have a natural tendency to selfishness. This implies the need for systems, or institutions, of some kind to help either develop virtues or at least prevent the pursuit of harmful selfish desires. This is particularly important as it applies to political leaders. Four examples that involve propaganda and deception in order to justify violence are then briefly outlined, followed by a passage from the *Analects* in which Confucius is confronted with an analogous situation. From this, I draw three lessons for keeping our leaders on the proper path. I argue that rather than view international tensions as signs of a clash of civilizations, we do better to turn again to the "Confucian solution" of seeking to improve *ourselves*, rather than correct the "other," and to hold *our own* leaders in check when sabers begin rattling.

To hold them in check, we must be informed ourselves. That involves the responsibility to critically investigate the facts put forward to justify violence or other morally dubious practices.

Many people have suggested that a unified world government would provide a solution to the problem of war. In the final chapter, I express skepticism toward this view. That chapter, "Two Visions of Confucian World Order," considers two competing interpretations of Confucian world order, labeled "Xunzian globalism" and "Mencian international harmony," arguing for the wisdom of the latter. It juxtaposes these two Confucian systems of world order with the Project for a New American Century, which was a neo-Conservative effort to dominate the world by leveraging American military strength. A genuinely Confucian project for world order, in contrast, would be an attempt to achieve a harmonious world by means of non-coercive ethical strategies. Once again, the Confucian solution is to "lead them with virtue."

THE DUTY OF REMONSTRANCE

As particularly emphasized in chapter 7, the early Confucians often faced inhumane and insincere rulers. The role of the ideal Confucian intellectual is to discern such duplicity and to remonstrate with authority figures who make immoral decisions. The obviousness and ubiquity of political deception notwithstanding, accusations of such deception, when applied to Western leaders, and particularly to *our own* leaders, are regularly dismissed as "conspiracy theories."[3] As we will see, Confucius himself played the role of conspiracy theorist (in the non-pejorative sense), at least once (*Analects* 16.1; see pp. 121–122), and Mencius did not shy away from speaking truth to power. For example, Mencius remonstrates with King Xuan of Qi, saying:

> How could it be acceptable [for you] to kill the older males, bind the younger ones, destroy their ancestral temples, and abscond with their valuable items? The whole world certainly fears the strength of the state of Qi. You have doubled your territory without putting *ren* (humane/benevolent) governance into practice. This will galvanize the armies of the world [against you]. If Your Majesty quickly issues orders to return their old and young, stop [looting] their valuable items, and, in consultation with the masses, establish a ruler and then leave them, then it is still possible that you may successfully stop [the coming attack against you]. (*Mencius* 1B11)

Although Confucius and Mencius traveled from state to state speaking to various rulers, the general Confucian advice regarding the responsibility of intellectuals is that they are to remonstrate with *their own* leadership. For example, Xunzi explains, "When a ruler is involved in schemes and affairs which go too far, and one fears they will endanger the state, high officials and senior advisors are able to approach and speak to the ruler. Approving when one's advice is used and leaving when it is not is called 'remonstrance'" (*Xunzi* 13/64/1; 13.2). Similarly, the *Book of Ritual Propriety* recommends discreet remonstration, and, "after remonstrating three times, if one's advice is ignored then one should flee" (*Li Ji,* Quli II). Such remonstrance can be thought of as a kind of loyalty to the very person one remonstrates with (or to the institution in which that person functions), for the immorality that this remonstrance seeks to avoid is self-undermining. As Xunzi explains, tyrants of the past met tragic ends—"Jie died on Mount Li, and Zhou's head was hung on a red pennant"—because "They could not foresee these events themselves, and there was no one who would remonstrate with them" (*Xunzi* 21/102/17; 21.2).

Selfishness and political duplicity are perennial problems. Especially when it comes to war and peace, it is important to help assure that our leaders stay on the proper path. The study of Mencius and Xunzi can help us think through what that path should be and how to stay on it.

NOTES

1. Xunzi writes, "The exemplary measure themselves with a stretched cord. In their contacts with others, they use a bow-frame. Because they measure themselves with a stretched cord, *they may be taken as a model worthy of emulation everywhere.* By using a bow-frame in their contact with others, they are thus able to be magnanimous and tolerant" (*Xunzi* 5/20/2; 5.7).

2. See Hagen 2007, 2019, 2010b, and the discussions of *li*, *yi*, and *dao* in Chapter 1, below.

3. There has been a fair amount of scholarly debate on the philosophy of conspiracy theories over the last two decades—to which I have contributed (Hagen 2010a, 2011a, 2018a, 2018b, 2018c, 2020a, and 2020b)—and the emerging consensus (which some of us thought was obvious from the beginning), is that each so-called conspiracy theory must be evaluated on its own merits. There have been a number of scholarly attempts to show that conspiracy theories, because they are conspiracy theories, can be safely dismissed. All have failed (see Coady 2006, 2007, and Hagen 2020b).

Chapter One

A Brief Overview of Confucianism

The classical period of Chinese philosophy begins with Confucius (Kongzi), who was born in 551 BCE. The second great Confucian thinker, who is said to have been a student of Confucius's grandson, was Mencius (Mengzi). The third was Xunzi, who came to prominence near the close of the classical period, which ended with the reunification of China in 221 BCE. My analysis will be based largely on the most reliable texts we have of the thought of these three thinkers: the *Analects* of Confucius, the *Mencius*, and the *Xunzi*, and primarily the latter two, since the *Analects* does not record extended discussions of war. This approach is in line with that of several other significant scholars who address Confucianism's relevance to contemporary political philosophy.[1]

Occasionally I will include consideration of two rival schools of thought, Mohism and Daoism. Mohism, founded by Mo Di (referred to here as "Mozi," that is, Master Mo), presented the first major philosophical challenge to Confucianism. The Mohists, who criticized Confucianism for its purported extravagance, advocated "impartial care" for all and strongly opposed aggression. The two most influential Daoist texts are the *Laozi* (also known as the *Daodejing*) and the *Zhuangzi*. Though there are various strands of thought expressed within these texts, generally speaking, Daoism favors a simpler, more natural and spontaneous way of life, compared to the structured and orderly life advocated by Confucians, who emphasize the importance of norms of ritual propriety.

Below I will briefly introduce the three key figures in early Confucianism: Confucius, Mencius, and Xunzi.[2] Then I will introduce seven defining elements of Confucian thought. Familiarity with the key figures and key concepts of Confucianism will enrich one's understanding of the following chapters.

CONFUCIUS (551–479 BCE)

Confucius strove to recreate the harmonious conditions that are imagined to have existed in the time of the semi-mythical sage kings. His strategy centered on moral leadership. According to Confucius, if influential people behaved as moral exemplars, they would inspire like conduct throughout the population. This was the simple key to achieving and sustaining a flourishing society. What was particularly important, then, was to convince rulers to behave well, and this involved convincing them to govern compassionately. But to get someone to govern compassionately, they would first have to *become* compassionate. And this requires a process of self-cultivation, for which ritualized norms of propriety play an important role. Focusing on the potential for moral development, Confucius offers a philosophy of hope and encouragement, as opposed to restriction based on fear, which would be a more legalist approach (such as that offered by the late Warring States philosopher Han Feizi).

MENCIUS (CA. 372–CA. 289 BCE)

Mencius is most famous for his position that people naturally have altruistic impulses. While Confucius had said, "In their natural dispositions, human beings are similar; it is by habit that they diverge" (*Analects* 17.2), he did not say whether these dispositions were good, bad, or neutral. Mencius, on the other hand, is unequivocal: "The goodness of natural human disposition is like the tendency of water to flow downward" (*Mencius* 6A2). To establish his position, Mencius provides a thought experiment: If, as a stranger passed by, a child was about to fall into a well, the stranger would be spontaneously moved to prevent the child's fall. Mencius is careful not to go so far as to say that everyone would actually *act* on this impulse. But the key point is that the impulse is there, even in the absence of any prospect of gain. The problem, in Mencius's view, is that although we all have what he calls "sprouts" of virtues, these sprouts can be crushed by environmental factors, the wrong sorts of experiences. So, despite the goodness of natural human dispositions, individuals can turn out bad. However, Mencius believed, given a normal, nurturing family environment, like good soil to a plant, these sprouts would naturally grow into the full virtues of *ren* 仁 (benevolence), *yi* 義 (appropriateness), *li* 禮 (ritual propriety), and *zhi* 智 (wisdom)—the first three of which will be described below.

In terms of political philosophy, Mencius tried to persuade rulers to exhibit the quality of *ren* (benevolence) and *yi* (appropriateness) in their governance. He argued that this would be in their own interest as well.

XUNZI (CA. 325–CA. 235 BCE)

While Mencius conceives of virtues as growing naturally, Xunzi regards them as artificial, devised by the wise exercise of intelligence, and developed with deliberate effort. Indeed, Xunzi believed that Mencius's position regarding the goodness of our natural dispositions is not only wrong but dangerous. If Mencius were right, Xunzi argues, there would be no need for the traditional norms of propriety (*li*). To emphasize his point, Xunzi repeatedly states that natural human dispositions are detestable. By this he really means merely that people have natural desires that will lead them into conflict, unless they are held in check, or appropriately channeled.

It seems to me that both Xunzi and Mencius have fair points. Xunzi, clearly, is right to point out that it is natural for humans to have self-centered desires, which, if unchecked, could lead people to quarrel. Mencius's point is perhaps a bit more contentious. Are our altruistic feelings natural, or learned? I am inclined to think that there is probably a natural *component* to them, at least. If so, we have a complex emotional nature that includes both good aspects and problematic (selfish) ones. It should be noted that Xunzi's emphasis on following ritualized norms of propriety, as a means to either reform or refine our dispositions, seems applicable even if Mencius is right about altruistic motivations being innate. Still, differences in degree may be important. Mencius worries that if we don't treat our positive qualities as fragile plants, we may injure them by working too hard to improve them, like the foolish farmer who pulls on his seedlings to help them grow (*Mencius* 2A2).

SEVEN KEY CONFUCIAN CONCEPTS: *JUNZI, DE, REN, LI, YI, TIAN,* AND *DAO.*

To get the most out of the following chapters, it will be helpful to have some background understanding of key Confucian concepts. So, I will here discuss seven such concepts, starting with five that are closely tied to the idea of moral leadership. They are *junzi* (exemplary persons), *de* (influential virtue), *ren* (benevolence), *li* (ritual propriety), and *yi* (appropriateness).[3]

Junzi 君子 (exemplary persons) and their *de* 德 (influential virtue)

"*Jun*" 君 means ruler, and "*zi*" 子 in this context means son.[4] So, originally, *junzi* referred to a prince. But Confucius used this term to mean someone who was *genuinely worthy* of the respect shown to a prince, that is, an "exemplary person." An exemplary person is someone whose conduct is worthy of emulation. And further, exemplary persons develop a quality called *de* 德 (influential virtue, or potency of character), which is a kind of moral charisma. They inspire others to act admirably. According to Confucians, it is the moral leadership of exemplary persons that makes the achievement and maintenance of orderly and harmonious life possible. Xunzi, Mencius, and Confucius, each in their own way, indicate the centrality of the influence of exemplary persons:

> Xunzi Writes: "Exemplary people (*junzi*) achieve the epitome of compelling character (*de* 德). Though silent, they serve as an analogy. Though not bestowing gifts, they are held dear. Though showing no anger, they are held in awe." (*Xunzi* 3/11/7; 3.9b)

> Mencius says: "Exemplary persons (*junzi*) initiate courses of action that bequeath order to posterity, enabling continuity." (*Mencius* 1B14)

> And, as we saw above, Confucius says: "An exemplary person's virtue (*de* 德) is the wind; a petty person's virtue is the grass. When the wind flows over grass, it is sure to bend." (*Analects* 12.19)

The three most distinctive characteristics of exemplary persons, which contribute to their *de* are: exhibiting *ren* (benevolence, or empathetic effort), observing *li* (norms of ritual propriety), and being sure to act with *yi* (appropriateness). I will discuss each in turn.

Ren 仁 (benevolence, empathetic effort, genuine-humanity)

Confucius said, "If 'exemplary persons' (*junzi* 君子) abandon *ren* 仁, how can they fulfill their title? Exemplary persons do not depart from *ren* even for the duration of a meal" (*Analects* 4.5). But what is *ren*? Often translated as "benevolence," *ren* is an enigmatic term, and not just because it is foreign to us. Confucius's students asked repeatedly for clarification of its meaning, and they received differing answers. In one instance, Confucius likened *ren* to "respect, magnanimity, trustworthiness, diligence, and kindness" (*Analects* 17.6). Elsewhere, when asked about *ren*, he remarked, "When dealing with the public, do so as though receiving important guests; in putting the people to work, do so as though performing a great sacrificial ritual. Do not press on others what you yourself do not desire, and you will incur no resentment in the state or in your family" (*Analects* 12.2).

Here is one useful way of thinking about it: *Ren,* which is associated with *ai* 愛 (love, care for others), can be thought of as a combination of two other Confucian concepts, *zhong* 忠 (doing one's utmost) and *shu* 恕 (empathetic consideration). In a word, it is *empathetic effort*. When applied to a ruler it implies compassionate governance, care for one's people as a parent for a child. But one must not forget how fundamental this is for Confucians. Exhibiting this consistent care for others is not just laudable, *it is the purpose of life*. As Mencius says, "To be *ren* 仁 is to be a person (*ren* 人)" (*Mencius* 7B16)—notice that the "person" radical occurs (on the left side) in the character for the virtue *ren*. Becoming a person of *ren*, developing the disposition to consistently care for others, is genuine self-actualization, and so, another gloss for *ren* could be "genuine humanity." Putting these ideas together: To have *ren* means to fully realize one's human potential by consistently working for the benefit of others, using imaginative introspection to understand what one ought to do.

According to Mencius, *ren* is a virtue that grows naturally out of innate "sprouts" of goodness. Moral development, for Mencius, is about allowing one's natural nascent altruism to grow, and thus to extend to greater and greater circles. Mencius said, "For all people there are things which are unbearable. *Ren* 仁 is extending this [attitude] to what one can bear" (*Mencius* 7B31).[5] For example, it is naturally difficult to bear witnessing the mistreatment of our loved ones, while bearing the mistreatment of distant others is not as difficult. Becoming *ren* is a process of extending our circle of care (though not to the point of *equal* care for all).

Li 禮 (ritualized norms of propriety, often rendered "ritual propriety")

Even if we aim to be *ren*, we may need guidance, and perhaps nudging. We need norms of propriety (*li* 禮) that set expectations and provide guidelines. One of Confucius's students, Zixia, remarks, "Exemplary persons (*junzi* 君子) are unfailingly reverent. In their engagements with others, they are respectful and observe ritual propriety (*li*)" (*Analects* 12.5). Confucianism's emphasis on *li* is perhaps its most distinctive feature. Originally, *li* referred to formal rituals, especially sacrificial rituals. But Confucius again used this word in a novel way. Although still applicable to formal rituals, for Confucius it had broader scope. Indeed, Confucius says, "Do not look in a way which is not *li*, do not listen in a way that is not *li*, do not speak in a way that is not *li*, do not move in a way that is not *li*" (*Analects* 12.1). But this does not mean life is scripted. Performing *li* skillfully bears a closer analogy to improvising in key, with attention to the passing chords, than it does to sight-reading. Attention to detail, and the meanings one is communicating in one's ritualistic actions is a vital component of *li*. *Li* provides a way of being in the world and relating to others. As Xunzi puts it, "[Exemplary persons] never leave [*li*]. For it is their sacred world and palace" (*Xunzi* 19/93/2; 19.3). To have *li* is to engage people and situations as though life itself were an elaborate ceremony. (As suggested in *Analects* 12.2, quoted above.) This way of being requires a sophisticated understanding of not only various formal and informal rituals, but also attention to the details of a situation and a developed sense of how to apply the various norms that govern behavior. Norms of ritual propriety (*li*) give guidance for conducting oneself in a way that both cultivates one's own character and facilitates productive and harmonious relations with others. As Joseph Chan explains:

> [*Li*] are not just a form of social etiquette but also perform important social functions—they help moral cultivation by regulating unhealthy desires and refining feelings and attitudes (*li jiao*, ritual education); they express the basic principles of human relationships and roles (*li zhi*, ritual system); these functions in turn help achieve a harmonious ethical society, which is the goal of Confucian governance (*li zhi*, ritual rule) that cannot be achieved by the use of penal law. (Chan 2014, p. 2)

Yi 義 (appropriateness)

Following norms is not enough. One needs a developed sense of appropriateness (*yi*) to complement *li*. If *li* were just an elaborate script, one wouldn't need *yi*. But life is too complicated and subtle for that. Navigating life productively requires not only the understanding of norms of ritual propriety, but also a developed internal moral compass that enables one to know when to approve and when to disapprove, and how to finesse a situation in the most socially constructive way. Because there may be more than one exemplary way to do that, *yi* is best understood under a rubric like "appropriateness," which encourages a "many good answers" way of thinking about it. To translate it "rightness" or "righteousness," on the other hand, may encourage a "one right answer" framework which could be misleading.[6] Confucius says, "In the exemplary person's (*junzi* 君子) view of the civilized world (*tianxia* 天下),[7] there is nothing that is [always] suitable, and nothing that never is. They favor what is appropriate (*yi* 義)" (*Analects* 4.10). "To me," Confucius says elsewhere, "nothing is simply permissible, and nothing is simply impermissible" (*Analects* 18.8). On the other hand, Mencius seems to suggest that there are things that are just not to be done (*Mencius* 7B31, 4B8 and 7A17), including killing an innocent person (*Mencius* 2A2). And Xunzi seems to concur (*Xunzi* 8.2, 11.1a).

Tian 天 (nature, the heavens, sometimes translated "Heaven")

Confucianism has a religious, or at least quasi-religious, aspect. Although Confucians do not worship a supernatural God, they do view humanity as situated in a profound and awe-inspiring context, *tiandi* 天地, the heavens (*tian* 天) and the earth (*di* 地). And it is particularly *tian* 天 that they regard with a sense of awe. Often translated as "Heaven," *tian* sometimes just means "the sky." In other contexts, as an abbreviation of *tiandi*, it means something like "nature" and implicitly includes the earth and its regularities. This is especially the case in the *Xunzi*. An older tradition personified *tian*, and there are a number of passages in the *Analects* in which Confucius evokes *tian* in an exclamation such as, "Oh! *Tian* has bereaved me!" (*Analects* 11.9). In the discussion of humanitarian intervention, we encounter the question: Who may use the military to punish another? The answer is: Only an "agent of *tian*" (*Mencius* 2B8; see p. 64, below).

Dao 道 (way, path, guiding discourse)

Finally, a few comments should be made about the well-known yet enigmatic term *dao* (alternatively spelled "*tao*"). Most English speakers are familiar with the idea that the Chinese word *dao* means "the Way." But what exactly does that mean for early Confucians? First of all, there are no articles (such as "a" and "the") in Chinese. So, to translate *dao* as "*the* Way" involves an interpretation of the context and understanding of the worldview in which the term is used. Unfortunately, many of the earliest Western interpretations of Confucianism were strongly influenced by Christian biases. And so, there was a questionable tendency to look for, and find, a kind of absolute in this concept. The practice of capitalizing it has the effect of reinforcing the reification of *dao* as some *Special Thing* that exists—perhaps *eternally* (as many translations of the Daoist text the *Daodejing* suggest). Although the *the* is sometimes appropriate, or at least reasonable, the combination of the *the* and the capitalization, along with influential early absolutist interpretations, contributes to the continuation of a misunderstanding of fundamental early Confucian assumptions.

For example, Edward Slingerland, author of a well-regarded translation of the *Analects*, offers the following explanation of *dao*, consistent with the Western tendency to assimilate the idea to Christian thought:

Confucius seems to have been the first to use this term [*dao*] in its full metaphysical sense. Referring literally to a physical path or road, *dao* also refers to a "way" of doing things, and in the *Analects* refers to *the* Way—the unique moral path that should be walked by any true human being, endorsed by Heaven and revealed to the early sage-kings. More concretely this "Way" is manifested in the ritual practices, music, and literature passed down from the Golden Age of the Zhou. (Slingerland 2003, p. xxii, emphasis in original).

David Hall and Roger Ames provide a contrasting view that avoids the misunderstanding:

Commentators upon the Confucian *Analects* often nominalize *dao*, explaining it as a preexisting ideal to which conformity is enjoined . . . We shall argue that to realize the *dao* is to experience, to interpret, and to influence the world in such a way as to reinforce, *and where appropriate extend*, a way of life established by one's cultural precursors. (Hall and Ames 1987, p. 227, emphasis added)

In other words, for Hall and Ames, *dao* does not designate a "*unique* moral path . . . endorsed by Heaven . . . [and] manifested in the [specific and unchanging] ritual practices . . . passed down from the Golden Age." Rather, the *dao* grows out of this specific past, emphasizing the best parts and adapting as deemed appropriate by cultivated *junzi*.

It is useful to orient oneself with specific revealing passages. For example, Confucius said, "People are able to broaden *dao*, it is not *dao* which broadens people" (*Analects* 15.29). And Xunzi wrote, "D*ao* is not the *dao* of *tian* 天, neither is it the *dao* of the earth. It is that by which the people are led; it is the *dao* of the *junzi*" (*Xunzi* 8/28/15; 8.3). Passages like these suggest that *dao*, far from being some kind of absolute *Special Thing*, is a "way" in a more down-to-earth sense. It is more like "a way of doing things." For example, Mencius says, critically, "When it comes to regulating funerals, Mohists take frugality as their *dao*" (*Mencius* 3A5), meaning he doesn't think the Mohist way of doing things is adequately respectful of the dead. In addition, since *dao* can also mean "to speak," or "doctrine," there are contexts when it means something like "a guiding discourse," whether or not it is a good one. For example, Mencius complains, "The *dao* (道) of Yang and Mo are repeated incessantly, while the *dao* of Confucius is not [often] expressed" (*Mencius* 3B9), meaning the teachings of Confucius are not as popular as those of Yangzi and Mozi. "*Dao*" can also be used as a verb, meaning "to guide," or as in the following: "Exemplary persons proceed (*dao* 道) according to what is regular, while petty people reckon what might be exploited" (*Xunzi* 17/81/1; 17.5; cf. 4.8).

Still, although the path the Confucian *dao* advocates following is a human construction, it is certainly not some *arbitrary* set of guidelines. It must meet certain pragmatic standards; especially, it must tend to contribute to harmonious social living, given the enduring conditions of nature (*tiandi*) and natural human dispositions. At the same time, there is a conservative assumption that exemplary models of the past should be taken as guiding norms. Thus, *dao*, as it proceeds into the future, is a pragmatic extension of the past, exhibiting both continuity and change, as well as reasonableness and contingency. Though there are constraints on change and contingency, *dao* is a long way from being an abstract unchanging absolute.

NOTES

1. For example: Daniel A. Bell, Tongdong Bai, and Yan Xuetong. I address the work of these three, as it applies to Confucian just war theory, in Chapter 4, "Mencius on War and Humanitarian Intervention." And also, Sumner Twiss and Johnathan Chan, whose work I address in chapter 6.

2. There are many ways to introduce oneself further to the key figures of early Confucianism. Brief introductions include Fingarette 1972, Ivanhoe 2000, and Goldin 2011. There are also many translations with informative introductions, such as Ames and Rosemont 1998 (*Analects*), Van Norden 2008 (*Mengzi*), Hutton 2014 (*Xunzi*), and Hagen and Coutinho 2018 (a sourcebook of early Chinese philosophy).

3. For descriptions of more classical Chinese concepts see "Key Philosophical Terms" in Hagen and Coutinho 2018, pp. 29-56, and "The Chinese Lexicon" in Ames and Rosemont 1998, pp. 45-65.

4. In other contexts, *zi* 子 can mean child. It is the same character that, when it serves as a suffix to names, means "master," as in: Kongzi (Master Kong, Confucius), Mengzi (Master Meng, Mencius), and Xunzi (Master Xun).

5. Cf. "The tender feeling we have toward our closest relatives is *ren* (仁). The respect we show toward our elders is appropriateness (*yi* 義). There is nothing else to do but make these manifest throughout the world" (*Mencius* 7A15).

6. Consider that it is not perfectly idiomatic to say "*the* appropriate way" or "*a* right way." Instead we tend to say "*an* appropriate way" (suggesting that there may be others alternatives) and "*the* right way" (suggesting exclusivity).

7. *Tianxia* 天下 refers to all below (*xia* 下) the heavens (*tian* 天), in other words, the known civilized world from the perspective of the early Chinese.

Chapter Two

Western and Chinese Attitudes Regarding Warfare

In this chapter, I describe and trace the implications of two pervasive and enduring Western attitudes toward warfare—one that romanticizes it, the other that moralizes it. I will then identify and describe an attitude toward warfare that sharply contrasts with these two Western attitudes and can serve as a critique of them. It is found expressed in the Chinese classic treatises on warfare, the *Sunzi* and *Sun Bin*, as well as Daoist, Mohist, and Confucian classics. I will also consider reasons to think that it would be imprudent for the West to abandon its romantic and moralistic attitudes. In the end, I introduce a Confucian-inspired long-term strategy for maintaining peace and security that could resolve a contradiction associated with the Western strategy of peace through strength. This "Confucian solution" will then be elaborated over the course of the succeeding four chapters.

The main contrast I want to highlight is between the Western romantic attitude toward war and the gloomy Chinese one, which I originally wrote about in the 1990s (Hagen 1996).[1] This contrast has been noted by other scholars, including Daniel A. Bell, Cao Qin, and Hans-Georg Moeller. A few of their observations are worth noting at the outset:

> This Confucian idea that war should be considered non-ideal, and unfortunate but occasionally necessary event stemming from difficult circumstances, may seem obvious today.... However, it contrasts with the historical glorification of warfare and the romantic model of the heroic soldier characterized in terms of boldness and masculinity that has been so prominent in Western societies. ... There would be no need for a Chinese critic of warfare to argue against the view that war is "charming" [as did Marcel Gauchet]. (Bell 2008, p. 248 n24)

> One feature of traditional Chinese political thought was that it rarely glorified military virtue.... The case was rather different in Western and other traditions.

> From Homer's description of heroic Greek warriors to the fascist worship of war in the twentieth century, there has been a trend (though usually not the mainstream one)[2] in Western tradition to see military activities as honorable and praiseworthy in themselves.[3] (Qin 2020, p. 164)

> As opposed to many western images of war and warfare, the "defensive" *Laozi* does not connect war with heroism, justice, and collective pride. Since Greek antiquity, Western representations of war have been tightly connected to images of heroes and deeds of heroism—and this general trend continues uninterrupted into the Hollywood movies of today. . . . the *Laozi*, although very much concerned with matters of war, lacks such images. In the *Laozi*, the military leader appears, if seen from the perspective of heroism, more like a coward. He is, if at all possible, on the retreat; he shuns battle, and his major concern is to stay hidden as much as he can. (Moeller 2006, pp. 82-83)

While the contrast between the Western romanticism of war and the more negative Chinese attitude is clearest in the Daoist text *Laozi* (especially *Laozi* 31), as well as in the *Mozi* (especially chapter 18), it is perceivable also in the military treatises and Confucian classics as well. I will begin by describing the Western romantic attitude with which the Chinese attitude contrasts.

TWO WESTERN ATTITUDES

The Romance of War

There is a long-standing practice of romanticizing warfare in Western culture that we do not find in the classics of ancient China. Embedded deep in the Western tradition, and continuing to this day, is the underlying assumption that war is a masculine activity. Involvement in combat is a condition almost necessary for true masculinity, or at least the highest expression of it. This masculine character of war is fundamental to its romanticism and is connected to the concept of war as a duel.

Niccolò Machiavelli, in his *Art of War*, uses the term *virtù* to describe the quality of the good soldier. In Latin, "*vir*" means manliness, and for Machiavelli it implies boldness, audacity, aggressiveness, inner strength, resoluteness, potency, leadership, and self-aggrandizement, as well as other such masculine qualities; it signifies masculine self-realization. Neal Wood, in his introduction to the *Art of War*, writes, "Among the attributes included in *virtù* are boldness, bravery, resolution, and decisiveness. A tour de force, military or political, results from the vital creative energy so much a part of *virtù*. . . . [A]ll that characterizes a woman in Machiavelli's view is opposed to *virtù:* faintheartedness, irresolution, and hesitancy" (Wood 1965, p. lv).

Machiavelli contrasts *virtù* with *fortuna*, represented as feminine. She is natural circumstance, spiritual, incomprehensible, and paradoxically respects but resists *virtù*. And, while she does have a will, she may change her mind in response to *virtù*. However, though she may yield, she is not the weaker of the pair: "in battle, *fortuna* has often a much greater share than *virtù*" (Machiavelli 1965 [1520], p. 202).

Now consider Carl von Clausewitz's ideal general. As described in his treatise *On War*, he is not a man of broad virtues or education, as is the characteristic Chinese general,[4] but rather is narrowly focused on the practical skills required to execute a military campaign. "Only those activities emptying themselves directly into the sea of war have to be studied by him who is to conduct its operations" (Clausewitz 1968 [1832], p. 196). Clausewitz's military genius is someone with great force of will and strength of mind. He is not so different from Machiavelli's man of *virtù*. He is a man of great passion; courageous and resolute, he is always balanced and stable. The strong mind, Clausewitz explains:

> balances the excited passions without destroying them; and it is only through this equilibrium that the mastery of the understanding is secured. This counter poise is nothing but a sense of the dignity of man, that noblest pride, that deeply seated desire of the soul always to act as a being endued with understanding and reason. We may therefore say that a strong mind is one which does not lose its balance even under the most violent excitement. (Clausewitz 1968, p. 147)

Likewise, Clausewitz's concepts of "great moral courage" and "true military spirit" ring familiar with the sound of *virtù*. He describes an army imbued with the latter as follows:

> [A]n Army with all its physical powers, inured to privations and fatigue by exercise, like the muscles of an athlete; ... always reminded of its duties and virtues by the short catechism of one idea, namely the honour of its arms; Such an Army is imbued with the true military spirit. (Clausewitz 1968, p. 255)

Clausewitz asserts, "War is nothing but a duel on an extensive scale" (Clausewitz 1968, p. 101). We may consider the model of war as a duel both in terms of what transpires on the battlefield and on the home front, with the mentality of the soldier mediating between them. On the battlefront, war is a masculine activity characterized by boldness and courage—Machiavelli's *virtù*, Clausewitz's true military spirit. The home front instantiates everything that the war is to defend: one's country, home, way of life, and family (wife and mother)—just as a duel is often to defend one's honor. What connects the two is the mentality of the soldier—a romantic mentality involving the concepts of chivalry and gallantry.

While Machiavelli is often more concerned with *virtù* than virtue, he *is* concerned about the morality of the troops:

> But above all, we ought to have strict regard to their morals and behavior; otherwise we shall choose men who have neither modesty nor honesty, who will be a scandal to an army, and who not only become mutinous and ungovernable themselves, but sow the seeds of corruption among others. Is it to be expected that any kind of *virtù* or praiseworthy quality can be found in such men? (Machiavelli 1965, p. 34)

The virtue of the good soldier, where good implies both effectiveness and morality, is the "Officer and Gentlemen" model of chivalry that has continued to be integral to the Western romanticism of war.

By a "romantic" notion of war, I do not simply mean an idealistic but false view, but rather mean to imply that how the sexes relate to one another is involved. Consider that gallantry and chivalry connote both courage on the one hand and courtesy (especially toward women) on the other. Furthermore, if we are to press the metaphor of war as a duel, do not duels often revolve around women? And if this is the case, should we not expect the action that settles the dispute to also determine who is the most masculine? As Nel Noddings notes, "One element in the attraction to war is a long history of associating masculinity with the warrior . . . Not to exhibit a warrior's courage is to lose one's manliness" (Noddings 2006, p. 38).[5] Indeed, as John Ruskin, a prominent nineteenth century British author and social critic, puts it:

> [T]he great justification of this game is that it truly, when well played, determines *who is the best man*;—who is the highest bred, the most self-denying, the most fearless, the coolest of nerve, the swiftest of eye and hand. (Ruskin 1964 [1866], p. 45)

We test our mettle against theirs. Whoever's *virtù* is greatest is deemed not only stronger but also, therefore, better. In the Western tradition, gallantry, chivalry, and heroism are inseparable from the romantic attraction of fame and honor. Self-aggrandizement involves increasing the scope of one's prestige or honor. Clausewitz rhetorically asks, "Has there ever been a great Commander destitute of the love of honour, or is such a character even conceivable?" (Clausewitz 1968, p. 146) As for courage, it is not courageous to slaughter the helpless. For courage in warfare to make any sense, the opponent must present a real threat; they must be considered in some sense comparable. Only in the face of danger can one demonstrate his courage and general masculinity.

The romanticism of warfare has been part of Western tradition and literature since at least the time of Homer. Graphically depicted in the *Iliad*, war was a gory mess. What would drive so many men to take part in this endeavor? Achilles answers plainly, "[W]hy must the Argives make war on the Trojans? Why hath Atreides gathered his host and led them hither? Is it not for lovely-haired Helen's sake?" (Lang 1921, p. 171)

In Shakespeare's *Henry V*, the king's pre-battle speech argues that his men, though strategically disadvantaged, fatigued, and outnumbered, can prevail by shear masculine force. "The fewer men, the greater share of honour. God's will! I pray thee, wish not one man more." (Act IV Sc. III Lines 22-23). His speech succeeds in inspiring his men. "God's will! You and I alone . . . could fight this royal battle!" (Act IV Sc. III Lines 74-75). Their valor proves to be the greater as they proceed to carry the day, thus their manhood is affirmed as promised:

> And gentlemen in England now a-bed
> Shall think themselves accurs'd they were not here,
> And hold their manhood cheap whiles any speaks
> That fought with us upon Saint Crispin's day. (Act IV Sc. III Lines 64-67)

It should also be remembered that while the battle is the climax of this play, it is not the end. The end is a marriage. King Henry wins the battle, and from this he wins Kate's hand. As John H. Walter comments, "The Christian prince to complete his virtues must be married" (Walter 1954, p. xxviii).

In Ruskin's 1866 address to the Royal Military Academy, he asserts that the motivations for the young men joining the army are those of "sentimental schoolboys":

> Because, on the whole, it is the love of adventure, of excitement, of fine dress and of the pride of fame, all of which are sentimental motives, which chiefly make a boy like going into the Guards better than into a counting-house. (Ruskin 1964, p. 52)

Such an effort to undermine the romantic notion of war serves as confirmation of its pervasiveness.

Interestingly, Ruskin puts the responsibility for war on women. "Only by your command, or by your permission, can any contest take place among us" (Ruskin 1964, p. 60). If in fact, from the view of the males involved, war is romantic in the amorous sense, an attempt to please or be admired by females, then Ruskin is right in attributing to women the power to control war, and to end it. As long as the romantic view captivates their minds, men do not possess the determinative power. "Bid them be brave;—they will be brave for you; bid them be cowards;—and how noble soever they be, they will quail for you. . . . [S]uch and so absolute is your rule over them" (Ruskin 1964,

p. 60). If women truly wanted wars to end, Ruskin suggests, they could end them: "while any cruel war proceeds, she will wear black;—a mute's black, —with no jewel, no ornament, no excuse for, or evasion into, prettiness—I tell you again, no war would last a week" (Ruskin 1964, p. 61). The fact that women, in no fewer numbers than men, can be found not wearing black but waving flags, signals their complicity in the exercise of war. Indeed, just fourteen years after Ruskin's death, the British "white feather girls" of WWI publicly shamed young men into enlisting. If men will volunteer to face death to answer the charge of cowardice from women, perhaps they would be willing to abstain from participating in violence if that, instead, was what women expected of them. But if men can be excused from responsibility due to their blindness born of romance, so too can women, for they are presumably no less entranced. Romantic sentiments embedded deep in one's psyche, especially when one is unaware of them, skew one's choices.

We should not be too surprised that the warrior is judged in relation to masculine qualities. It has been predominantly males who have organized and engaged in these activities, and who have written about them. However, some may be tempted to say that modern warfare with its weapons of mass destruction has become impersonal, and that the romantic notion of war can no longer prevail. But it is all too easy to show that, on the contrary, it is alive and well. The popularity of Rambo and other supremely masculine characters ought to suffice to convince us that the romantic attitude regarding war continues. Certainly, anti-war movies have been popular too, "Platoon" and "Born on the Fourth of July," for example. But efforts to de-romanticize war in the West have been peripheral, while romanticizing has been the central theme. The United States, to maintain its voluntary recruitment system for the military, uses advertisements that make blatant appeals to chivalry. Knights on a chessboard, for example, redressing an offense symbolize the Marines, characterized by the slogan "the few, the proud."

The romantic conception glorifies war and makes it seem a noble venture. It makes waging war more palatable than if it were thought of as only tragic and terrible. Thus, it seems that this conception of war would contribute to the propensity to make war. If so, does it follow that, in the name of peace, Western culture ought to dispel from itself this romantic notion?

Before we reject the romantic image, we should consider not only how it affects the willingness to go to war, but also how it affects the way war is conducted. Returning to Clausewitz's comment that warfare is a duel: in a duel there are rules that restrict the domain of action. And the contestants must be of comparable strengths to enable the demonstration of courage. These rules, and rough parity among participants, make a duel "fair," legitimizing the outcome. Whoever wins is "right," for they won fair and square. Further, the rules keep the violence within a "legitimate" scope, which is necessary for a duel to remain a noble affair. After all, if nobility implies courtesy, especially toward women, then "noble" acts of violence must not impact innocent bystanders, especially women and their children. Therefore, when this model is applied to war, some notions of fairness in the conduct of war are to be expected, and the massacre of innocents is prohibited, no matter what advantage such an action might provide one's side.

To abandon the romantic notion while still engaging in warfare risks the fierceness required of battle being untempered by the fairness required of chivalry, gallantry, and the accumulation of nobility and honor. The full and terrible elements of masculinity may manifest themselves unchecked—manly fierceness slipping into unbridled ferocity. If so, any atrocity that expediently achieves one's purposes may pass as acceptable. Acts we would find shockingly repugnant, when viewed directly in their nakedness, may in the heat of hostilities be perpetrated and condoned, or at least excused, if not constrained by a romantic ideal of moral fairness in war.

We know that moral restraint in war is fragile, for even with the West's romantic sensibilities intact, Western countries have perpetrated abominations: Atomic blasts, carpet-bombing, as well as village massacres. Still, these atrocities cannot go on too long; upon hearing of such incidents there is usually a public outcry, as the contradiction between our sense of moral limits and the inhumanity of our actual conduct becomes too stark.

Consider the curious timing of the end of the Persian Gulf War in 1991. Just when the American-led forces were in position to cut off and destroy Iraq's Republican Guard, the President, in consultation with his generals, chose to cease hostilities. That this decision followed the bombing of an Iraqi convey in retreat from Kuwait on what has come to be known as the "Highway of Death" is perhaps no coincidence. There was at once a fear that the public support for the war would be undermined by images of the carnage, as well as a feeling that to continue such a one-sided campaign would be to engage not in war but slaughter. General Powell is said to have remarked that to continue would be "unchivalrous" or "ungallant" (Liivoja 2013, p. 144).

Duties and Moral Limits

The romantic attitude toward war is not the only relevant influence on the Western psyche. For example, there is an influential legal tradition in the West that presupposes that moral limits must be maintained even in war. After a war there is often a strong inclination to punish those involved in "war crimes," a term that implies limitations to the scope of permissible actions in warfare. The Nuremberg trials after World War II present a particularly clear example. And yet it must be acknowledged that the victors generally excuse themselves for their own excesses, appealing to transcendent justifications or consequentialist ones, such as the explanation that the use of nuclear weapons hastened the end of the war and thus, on balance, saved lives.

At the same time, a particularly significant competing influence on the Western mind is the belief in a duty to fight. This involves a sense of rightness that moves people to believe that *not* to engage in war is to be morally remiss. For if a war is truly a *just* war, its fundamental justification, whether holy or humanitarian, lies in the protection of what is good against what is evil. I'm not so much referring to the nuances associated with the intellectual tradition known as just war theory, but to the raw emotion associated with those brought up to feel a duty to country and impelled by a visceral sense of honor.

Western soldiers go to war with both a determination to demonstrate their individual virtue (though perhaps it is really *virtù*) and a sense of righteousness in the cause. The first is a romantic, war-as-duel, motivation, while the second involves transcendently grounded justification. Superficially these seem to go together, for to have virtue is to defend goodness—a *heroic* defense of a *righteous* way of life. "A soldier's vow to his country is that he will die for the guardianship of her domestic virtue, of her righteous laws, and of her any-way challenged or endangered honour" (Ruskin 1964, p. 55).[6] But when soldiers depart to gallantly defend righteousness, the romantic and transcendent models often clash. For virtue can rarely be maintained in the execution of the task.

On the home front, there are principles that must be defended. They may include principles governing a way of life. Or, in the case of national defense, the principle may be that one must not allow one's nation or culture to be destroyed. On the battlefront, consequently, the overriding concern is to *win*. From this perspective, battle is not to *determine* who is best (and therefore right), but to *enforce* what one already knows is right. The mentality that links the home front with the war front is the certainty of one's moral correctness. Here, war is motivated by moral outrage.[7] We are good; they are evil. War is therefore a struggle to ensure that righteousness prevails.

One problem with this model, of course, is that we may not be entirely right. Our adversaries surely do not think we are. But perhaps the greater problem is in the conflict between the "must win" posture that we are forced into and the possibility, even likelihood, that in order to win we must become the very evil that we seek to defend against. That is, it is problematic to say, "We must defeat this evil at any cost, even if it means becoming evil ourselves." On the other hand, if we allow moral scruples to cost us victory, we will have permitted evil to dominate, which is unacceptable.

Although the romantic and moralistic attitudes have conflicting implications, they nevertheless dwell together in the Western psyche. Sometimes one takes precedence, sometimes the other. The United States answered, "Whatever it takes!" when it bombed Hiroshima, and when it firebombed much of Japan before that. When we accept this implication of the attitude of transcendent justification, it turns out that we are no longer acting consistent with the model of a romantic duel. The fundamental principle of following certain rules, including not killing civilians, is shattered, and all the nobility and chivalry there may have been is gone. We participate in an inglorious contradiction. Not only is the nobility of our conduct compromised, if our transcendent justification was a humanitarian one regarding basic human rights, then we do terrible violence to our own principle of justification.

THE CHINESE ALTERNATIVE

The Gloom of War

The Western romantic model characterizes soldiers in masculine self-aggrandizing terms such as boldness, courage, resolution, and potency. These soldierly qualities, so central to the Western romance of war, are not to be found as prominent concerns in the Chinese tradition. Consider bravery, and the risk that is necessarily associated with it. Western culture celebrates risk. However, as Hans-Georg Moeller points out, there is no "fearless, risk-taking fighter" depicted in the Daoist classic the *Laozi* (also known as the *Daodejing*); there is no "desperado," no "'lethal weapon'-type" (Moeller 2006, p. 83). In modern Western literature, such as the writings of Ernest Hemingway, Moeller points out, we find "the passionate fighter, Gen. George S. Patton type, a man whose whole (masculine) emotionality is focused on war and triumph in battle" (Moeller 2006, p. 83). However, Moeller notes, "[T]here is nothing like this in the *Laozi*; the element of passion (male) ecstasy is practically absent from its depiction of war" (Moeller 2006, p. 83).[8] This type is likewise absent in Confucian and Mohist discussions of war, as well as the Chinese military classics. Instead, for example, Confucius suggests that the issue of war should be treated with caution (*Analects* 7.13) and trepidation (*Analects* 7.11).[9] In fact, Confucians suggest that peaceful influence, rather than physical force, is the way to achieve the highest status and acclaim. Mencius said, for example, "Those who use force as a substitute for *ren* 仁 (humanity) are [mere] hegemons. . . . Those who use their influential virtue (*de* 德) and put *ren* (humanity) into practice are kingly" (*Mencius* 2A3). Trying to achieve status through war is like "climbing a tree to catch a fish" (*Mencius* 1A7; see p. 69) except that one is catastrophic and the other is merely ineffectual. At the same time, excelling at war is a "tremendous crime" (*Mencius* 7B4; see p. 79 n9) deserving of a punishment worse than death (*Mencius* 4A14; see p. 65).

In the Chinese military tradition, to take a risk is to be remiss. A battle is to be engaged not to test our skills or determine who is the braver or manlier, but to take what is, by virtue of one's position, a sure thing (see Ames 1993, p. 93). In the Chinese classics, we find neither glorification of war nor an emphasis on masculinity. As D.C. Lau and Roger Ames point out, "[I]n neither *Xunzi* nor *Laozi* is warfare 'gendered.' In the absence of a focus on egoistic agency, notions such as honor and glory are relatively uncelebrated, and instead the full range of possible gender traits are embraced as resources for dealing with warfare effectively" (Lau and Ames 1996, p. 110).

As for self-aggrandizement, this is strongly criticized in Daoism. So as not to "harm oneself," Zhuangzi (chapter 7) enigmatically warns against endeavoring to win and maintain fame, writing: "Do not be an embodier of fame," as Burton Watson translates it (Watson 1964, p. 94). More literally, the passage says, "Do not be a *corpse* (*shi* 尸) of fame,"[10] which suggests that the pursuit of fame is deadly. Elsewhere Zhuangzi explains, "Virtue is undermined by fame," describing fame as an "ominous tool" with which people "crush each other" (*Zhuangzi*, chapter 4). Similarly, the *Laozi* indicates that showing off results in calamity (*Laozi* 9),[11] and that "Having the courage to be heroic results in death; having the courage to not be heroic results in life" (*Laozi* 73). The ideal, the latter passage also suggests, is to be "adept at winning without contending."

The concept of winning without contending is echoed in the early Chinese military texts. According to the Sun Bin, "[T]he battle of the expert [does not] win him reputation for wisdom or credit for courage" (Ames 1993, p. 116). As discussed in more detail below, Sunzi, in the *Art of War*, writes, "[T]he expert in using the military subdues the enemy's forces without going to battle" (Ames 1993, p. 111). Ping-cheung Lo explains, "Sunzi's overall advice is that by means of creative non-violent combat, one can outwit one's opponents without actual military violence and win without destroying enemy lives and properties but rather preserving them. This is the pinnacle of military excellence" (Lo 2015c, p. 74).

As in the *Daodejing*, there is a palpably negative tone regarding warfare in the early Chinese military texts. P. J. Ivanhoe explains, "*Master Sun's Art of War* retains the general Chinese tendency to regard war as an unfortunate departure from life's proper work" (Ivanhoe 2011, p. xxv). "For Sunzi, war is to peace as illness is to health. . . . Sunzi never glorifies war or revels even in the prospect of victory,[12] and he is poignantly aware of how harmful war is to the welfare of states and the people within them. In these respects, *Master*

Sun's Art of War is like the *Daodejing*" (Ivanhoe 2011, p. xv). Sunzi's *Art of War* is typical of early Chinese military texts in this respect. Ping-cheung Lo explains, "[T]he just war discourses in *The Seven Military Classics* are often prefaced with the confession that 'weapons are ominous or brutal instruments.' No matter how righteous or just the war is, it is always tragic to wage war" (Lo 2015b, p. 35).

According to the Chinese classics on warfare, war is no way to win fame. There is no opportunity for self-aggrandizement here. Sunzi states:

> [A] commander who advances without any thought of winning personal fame and withdraws in spite of certain punishment, whose only concern is to protect his people and promote the interest of his ruler, is the nation's treasure. (Ames 1993, p. 150)

Sun Bin concurs, "[O]ne who takes pleasure in military affairs shall ultimately perish, while one who seeks to profit from victory shall incur disgrace" (Ames 1993, p. 87).

There is a famous Chinese story in which, over the objections and pleadings of King Helu of Wu, general Sunzi has the King's two favorite concubines beheaded (see Ames 1993, pp. 32-34). The point is to impress upon the King that "while [a general] is in command of the troops, [he] is not bound by [the King's] orders" (Ames 1993, p. 34). Though not the main point of the "lesson," one effect is to disassociate war and romance. In other words, if we look for amorous implications, war leads to loss, not gain. Daoism, Confucianism, and Mohism also stress the loss associated with war, taking into consideration the loss of life on both sides. The *Laozi* likens celebrating victory in war to taking pleasure in massacre (*Laozi* 31, see below). Xunzi explains how deaths on either side make warmakers hated, which leads ultimately to their ruin (*Xunzi* 9.7; see p. 85, below). And the *Mozi* provides a long list of misfortunes associated with war, including civilians who end up dying of starvation.[13]

Of the three chapters in the *Mozi* titled "Against Military Aggression," the second one (chapter 18) is all about loss. It argues, as Paul Van Els notes, "(1) loss outweighs gain, (2) losers outnumber winners, and (3) even winners eventually become losers" (Van Els 2013, p. 79). As Hui-chieh Loy explains, "War, the Mohists argue, is too costly even for the aggressor, too uncertain a way for them to achieve their real goals,[14] and, even when used successfully, poses dangers of its own to the victor" (Loy 2015, p. 235). On Mozi's view, even "Heaven" loses whenever there is a war:

> If all the people in the world belong to Heaven, as Mozi claims, then war implies that one group of Heaven's subjects (the inhabitants of one state) attacks another

group of Heaven's subjects (the inhabitants of another state). Whichever state wins, Heaven always loses due to the casualties on both sides. (Van Els 2013, p. 89)

Whether we search the military texts of Sunzi and Sun Bin, or Daoist, Mohist, and Confucian classics, we do not find a romantic conception of war as a noble duel in the ancient classics of China.[15] This commonality between the military texts and broader philosophical works ought not be surprising when we consider that "the militarist texts, rather than articulating a philosophical program of their own origin, most often represent an extension and instantiation of the philosophical tenets of the 'schools' to which they belong" (Lau and Ames 1996, p. 62). In reading the *Sunzi* (the *Art of War*) in particular, one notices Daoist influence in the emphasis on "formlessness" and movement that always answers the enemy's movement with its opposite, as opposed to a clash of strength against strength. For example, Sunzi advises:

> Just as the flow of water avoids high ground and rushes to the lowest point, so on the path to victory avoid the enemy's strong points and strike where he is weak. As water varies its flow according to the fall of the land, so an army varies its method of gaining victory according to the enemy. (Ames 1993, p. 127)

In addition, sentences like: "Disorder is born of order; cowardice from courage; weakness from strength" (Ames 1993, p. 120) could pass as being from a lost chapter of the *Daodejing*. Only when the passage continues do we see the distinctly military application: "The line between disorder and order lies in logistics; between cowardice and courage in strategic advantage; and between weakness and strength, in strategic positioning."

Here, however, I am concerned mainly with a common thread: the sharp criticism of an emotional inclination toward warfare. Sun Bin states, "a distaste for war is the kingly military instrument" (Lau and Ames 1996, p. 149). Unlike efforts in the West to de-romanticize war, in classical China we find these sentiments in central texts, not on the fringe as a critique of the center. For example, the *Laozi* states:

> Fine weapons are inauspicious, detested by all. Thus, those who have the way (*dao*) have nothing to do with them.[16] . . . Victory [in war] is not admirable. Those who admire it take pleasure in the massacre of people. Those who take pleasure in the massacre of people will, for this reason, fail to achieve their aspirations in the world. . . . When people are killed in large numbers, this is an occasion of great sorrow, grief, and wailing. Victory in war, therefore, is to be dealt with according to the rites of mourning. (*Laozi* 31)[17]

War, in the *Laozi*, is "not a matter of collective pride" (Moeller 2006, p. 84).[18] It is quite the opposite, as the passage above dramatizes. The *Laozi* regards war, in Moeller's words, as a "political malfunction" (p. 85) and a "social disaster" (p. 84) that "lacks all human grandeur" (p. 85). It is "a case of social disease, an 'inflammation of the state,' so to speak. . . . War can thus be understood as the most extreme symptom of general social or political disorder. . . . War is, in other words, a harmful eruption of desires" (p. 76). Indeed, Moeller emphasizes, "War is human desire in action. Ancient Daoist philosophy shudders at such an intensity of desires, and at the social disorder and the heightened activity that comes with it and is so clearly destructive. . . . [War is] the very worst type of human activism" (p. 77).

Although some aspects of the Daoist perspective on war are particular to Daoism, such as those connected to its negative view of desire and of activism, other aspects are pervasive in the early Chinese literature on warfare. Referring to the Daoist notion of the "efficacy of not fighting," Moeller comments:

> [T]he military strategy advocated in the *Laozi* can well be understood as a "mainstream" position. Within the larger context of the ancient Chinese reflections on warfare, it does not present a unique Daoist perspective. Instead, it represents a view that seems to have been rather generally held at the time and was in line with the then dominating semantics of war. (Moeller 2006, p. 81)

Moeller notes that, despite Napoleon's disastrous advance into Russia, which so dramatically demonstrated the potential efficacy of nonengagement, this strategy "did not enjoy such popularity in western traditions" (p. 82).

In the early Chinese view, there is generally an overall negative utility in engaging in war, even if you win. The *Laozi* repeatedly emphasizes, Ellen Zhang observes, that "nothing is more catastrophic than war, since it brings about killing, mutilation, and widespread devastation" (Zhang 2015a, p. 186), and thus, the *Laozi* suggests that (in Zhang's words) "the prosecution of the war is always of a losing battle" (Zhang 2015a, p. 192). The military classics concur. The *Sun Bin* states, "[E]ven ten victories out of ten, while evidencing an able commander, is still a source of national misfortune" (Ames 1993, p. 85). One should therefore avoid war if at all possible. The *Sun Bin* insists, as the *Laozi* implies,[19] "You must go to war only when there is no alternative" (Lau and Ames 1996, p. 153; Ames 1993, p. 85). And, if you must go to war then it is best if you can win without a battle. The *Sunzi* tells us: "[T]he expert in using the military subdues the enemy's forces without going to battle" (Ames 1993, p. 111). And furthermore, one tries to wage the least costly battle:

It is best to keep one's own state intact; to crush the enemy's state is only a second best. It is best to keep one's own army, battalion, company, or five-man squad intact; to crush the enemy's army, battalion, company, or five-man squad is only a second best. So to win a hundred victories in a hundred battles is not the highest excellence; the highest excellence is to subdue the enemy's army without fighting at all.

Therefore, the best military policy is to attack strategies; the next to attack alliances; the next to attack soldiers; and the worst to assault walled cities. (Ames 1993, p. 111)

This contrasts sharply with Clausewitz's characterization of the goal of war, which is to render the enemy's army helpless. According to Clausewitz, "The destruction of the enemy's fighting power is . . . always the means to attain the object of the combat" (Clausewitz 1968, p. 131). This contrast supports B. H. Liddell Hart's comment in the "Forward" to *Sun Tzu: The Art of War*:

Civilization might have been spared much of the damage suffered in the world wars of this [the twentieth] century if the influence of Clausewitz's monumental tome *On War*, which moulded European military thought in the era preceding the First World War, had been blended with and balanced by a knowledge of Sun Tzu's [Sunzi's] exposition on "The Art of War." (Hart 1963, p. v)

Chinese Moral Restraints

In the introduction to his translation of the *Sunzi*, Roger Ames argues that the early Chinese worldview makes no appeal to an unchanging Reality behind the changes in the world of common experience (see Ames 1993, pp. 43-63). Such a Reality is the natural home of fixed transcendent principles, assumed to be as absolute as divine commandments, even if not thought of consciously in religious terms. Platonic Forms are an example. The power of transcendent principles is in the (sometimes misplaced) moral confidence that they encourage. Such principles are not tied to place or circumstance, and thus can be fought for in distant lands without the messy business of considering all the local particulars.

Some evidence supporting the idea that transcendent principles have not played a major role in Chinese thinking can perhaps be discerned from the kind of warfare in which the Chinese have historically been involved: border warfare and internal disputes; they never sent armies overseas to fight for abstract causes. According to Rebecca Byrne, "Moral argument is not used so frequently in China as it has been in the West as a way of arousing the people to angry attack" (Stroble 1998, p. 177, citing Byrne 1974). Rather than an opportunity for a glorious adventure backed by transcendent moral convictions, war was a local misfortune.

One may worry that a lack of transcendent principles would make a "principled war," in which moral limitations are observed, at least unlikely. But it is easy to make too much of this point. Concrete norms of proper conduct and an empirically grounded sense of appropriateness can do much of the work of transcendent moral principles. Xunzi argues, in characteristic Confucian fashion, that successful commanders are humane and honorable, and that, in particular, norms of ritual propriety (*li* 禮) "serve as guidelines for achievement and fame." He adds emphatically, "Solid armor and sharp weapons are not sufficient to be victorious. High walls and deep moats are not sufficient to be secure. Strict commands and numerous punishments are not sufficient to be awe-inspiring. If one follows the way (*dao* 道) [of propriety], one will progress. If one does not, one will fail" (*Xunzi* 15/72/9; 15.4). Xunzi is also explicit in his prohibition of military deceit: "Actions such as attacking for the purpose of taking by force, and conducting deceptive movements, are the business of [unscrupulous] feudal lords. A person of genuine humanity (*ren* 仁), in employing the military, does not permit deception" (*Xunzi* 15/68/12; 15.1b).[20] It is not that deception violates a transcendent principle. Rather, it conflicts with norms that are important for becoming virtuous persons. Further, as I will argue in chapters 4 through 6, both Mencius and Xunzi believed that if one cannot secure a nearly bloodless victory by marching openly in the light of day, then one does not have sufficient support, and thus the moral authority, to justify military action. In any case, the essential point here is just that there are mechanisms in the Chinese tradition that can do much of the work of transcendent principles.

However, it may be thought, such non-transcendent restraints are weaker than transcendent principles, and are more susceptible to exceptions. Indeed, some people worry that without the backing of a "transcendent reality" moral rules are like mere contracts that "can be revised, revoked, altered, and amended at will" (Goodman 2001, p. 375). In contrast, transcendent principles are (ostensibly) fixed, as if in the stars. It is harder to talk one's way around them. As can be seen in the above quotations, Confucian constraints are grounded in prudence and efficacy. They are advocated because they are deemed to be effective at achieving their goal. Arguably, if the circumstances were such that other measures would achieve the desired results, then those measures may instead be adopted. The contextual nature of traditional Chinese restraints makes them appear fragile. We can never completely remove restraints from the context with which they are most closely associated, nor conclude with confidence that they are meant to hold in all circumstances. Perhaps abuse of highly context-sensitive standards such as appropriateness (*yi* 義), norms of propriety (*li* 禮), and humanity (*ren* 仁) by less than sage-like leaders, at least in part, accounts for the history of brutality in Chinese warfare. On the other hand, Mencius and Xunzi have complex reasons for reluctance to deviate from established norms, as will be discussed in the next chapter. In any case, historically, neither China nor the West have lived up to the demands set by their respective philosophies. But imperfect restraints are far better than no restraints at all. Cultural attitudes set limits on what even a tyrant can ultimately get away with.

PROBLEMS AND SOLUTIONS

Problems

One problem with using war to resolve conflicts is that little effort is put into maximizing the *overall* results. It is wasteful. War has been pursued by Western powers even when there was much to lose and precious little, if anything, to gain *for the whole* (think of WWI, for example). War, in this sense, is individualistic, but on a national scale. An individual nation with a sense of national dignity, not bending an inch in the face of insult or injury, will take up arms in pure indignation, concerned more with its own national honor than achieving the best outcome for all parties. But a harmonious outcome is precisely what Confucians, in particular, are concerned about—making the best of the circumstances. While individuals who are seen as the cause of disharmony are dealt with, the focus, from a Confucian perspective, is on achieving, maintaining, or restoring a practical and fitting organization of the whole.

In contrast to European wars, warfare in China was traditionally fought (at least ostensibly) on behalf of the people on *both* sides, not just the people on one's own side.[21] Europeans identified themselves more strongly with their individual states than with Europe as a whole. In China, the reverse was true. Therefore, the Chinese combatants, more than their European counterparts, identified with the opposing army, their similarities being more salient than their differences. This ultimately resulted in fundamentally different attitudes toward warfare. For the Chinese case, Roger Ames explains:

> [W]hat makes any military action "appropriate" and "proper" (*yi*) as opposed to "self-seeking" . . . is the claim that it serves the quality of the sociopolitical order as a whole rather than any particular interest group within it. Those persons promoting military engagement must make their argument[22] on the necessity of such action to revive and reshape the shared world order. (Ames 1993, pp. 68-69)

Making related points, Xinzhong Yao writes:

[I]n order to win over the hearts of the people, each side of a war would first claim its cause to be just and benevolent, making the world know that it was going to carry out the Way on behalf of Heaven (*ti tian xing dao*) to relieve the people and punish the guilty (*diao min fa zui*). On the one hand this ritual-like preparation makes war a tool for ethical purposes, and the process of war is therefore disciplined by ethical norms, by which the cruelty of war is to some extent confined or reduced. On the other hand it has significantly increased the hypocrisy of war, making use of moral justification as the fig leaf of one's ambition to conquer others, and to cover up the misery and suffering war is going to cause or has caused. (Yao 2004, p. 106)

Concern for ritual propriety (*li* 禮), appropriateness (*yi* 義), and benevolence (*ren* 仁) might be thought to have some civilizing effect, though it is not clear how much. Likewise, it is not clear how effectively Western romantic attitudes restrain soldiers and their leaders. Still, one may legitimately worry that for the West to give up its romantic mentality risks losing a safety net that made wars civilized to some extent.

And yet, even the combination of romantic sentiments and transcendent principles too often fail to prevent massacres, wanton destruction, and other atrocities. For instance, under the pressures of the Vietnam war, whatever moral qualms there might have been regarding the torture and assassination of civilians, they proved insufficient to prevent the atrocities perpetrated under the Phoenix Program, which implemented such tactics on a large scale (Valentine 1990).

It seems that the only way to prevent heinous war crimes may be to refrain from war altogether. Yet what about the need to defend against the aggression of others? Surely to sit back and allow Hitler to achieve world domination in World War II would be remiss.[23] At this point in history, we cannot responsibly lay down our arms—or at least that is what many people fear, and understandably so.

The attempt to maintain peace while retaining the romantic conception of war results in the policy of "peace through strength"—*masculine* strength. However, this strategy involves a means-ends tension that undermines its ability to assure peace. One cannot just build bombs and rest easy knowing that they will never be used. For a state's strength to be a deterrent, there must be a credible threat that the leaders are stalwart enough to use it. And thus the "peace through strength" model necessitates continuous cultivation of the warrior mentality. And from this fact "peace through strength" defeats itself, for it simultaneously cultivates strength and fosters war.

The Confucian Solution

In the following chapters, I will be critiquing attempts to read early Confucians as just war theorists. For our present purposes it may suffice to note that, as Richard Jackson has maintained, "For the most part, JWT [just war theory] assumes that the main source of state power lies in military strength" (Jackson 2020, p. 50). In contrast, the Confucian solution to the problem of "peace through strength" is to make a distinction between military strength and strength through benevolent government. Xunzi, for example, writes, "One who uses virtue (*de* 德) to unite people will be kingly. One who uses physical force will be weakened" (*Xunzi* 15/74/16; 15.6a). And so, while peace through military strength defeats itself, peace through moral strength does not. But how does peace through moral strength work? In the context of ancient China, it was viewed this way: The people under the rule of a benevolent king, who loves and cares for them, will support the king. Not only that, being adequately fed and not over-worked they will be physically strong. Further, the benevolent king, also being wise, is somewhat frugal in his own expenditures. He makes sure that grain is stockpiled so that the people will be able to weather unforeseen natural calamities. In short, the state of a benevolent king is strong in a variety of non-military senses: it is economically strong and secure, its people are healthy, their morale is high, and their sense of allegiance is robust. This supports defensive military preparedness as well.

The opposite is true of a tyrant. Xunzi writes, "Regarding relations between ruler and ministers, and between superiors and subordinates: Others exhibit oppression and cruelty, and mutual estrangement and resentfulness increase daily. I exhibit generosity and kindness, and mutual affection and care increase daily. Because of this, I can simply wait for my adversary to collapse of his own accord" (*Xunzi* 9/41/24; 9.19b). This is an example of the Confucian view that there is a rough coincidence between what is appropriate and what is prudent. Xunzi elsewhere writes, "If superiors do not elevate ritual propriety, [the state] will be militarily weak. If superiors do not care for the common people, [the state] will be militarily weak. If promises are not lived up to, [the state] will be militarily weak" (*Xunzi* 10/47/19; 10.13).

We can see a parallel in current affairs. The United States military has been stretched thin in its muscular engagements in Iraq and Afghanistan. Economically, these misadventures have contributed to a potentially dangerous level of debt. Further, since the United States is not deemed trustworthy in its intentions, its "welcome" as a liberator has been short-lived, and thus conflicts that deplete resources and morale drag on and spread. Now consider a Confucian alternative. Instead of invading Afghanistan and Iraq, the U.S. could have invested in rebuilding its own infrastructure, thus providing both jobs and a foundation for its economic future. It could have, at the same time, provided sensible benevolent assistance to various peoples in need around the world (food assistance, vitamins, medicine, and small-scale infrastructure and such). America could significantly improve the conditions of hundreds of millions of people around the world, as well as domestically. If the United States had simply taken the high road, would its people not be safer? Economically it would have been stronger, and the economy ultimately supports a military. Further, the U.S. would have more friends around the world, and fewer enemies. Although perhaps fewer terrorists would have been killed, fewer people would have been driven to terrorism to begin with, and the rationale and motivation for terrorists to strike the U.S. would have been reduced considerably. In the end, it would have been better to adopt the Confucian solution.

In the next chapter, I address a number of considerations involving Mohism and Daoism as well as Confucianism which ultimately support my contention (explicitly argued for in chapters 4 through 6) that early Confucians had reasons to balk at supporting wars even if they have plausible humanitarian rationales. For example, I argue that even when norms of propriety appear to conflict with the achievement of desirable goals, early Confucians have reasons for sticking to the way of propriety, rather than making exceptions when hard pressed. These considerations bolster the suggestion, made in the present chapter, that Confucian norms can serve as effective restraints. They also serve as background for my argument that early Confucians would, and contemporary Confucians should, resist the temptation to support wars even when they are (ostensibly) based on humanitarian considerations.

NOTES

1. The present chapter is a revised and extended version of that article, namely, Hagen 1996.

2. If what is meant by "military activities" is soldiers serving in wartime, treating this as intrinsically honorable and praiseworthy is certainly mainstream in America, almost coercively so. Such an attitude is implied in our national anthem; people are expected to stand respectfully at major sporting events as they listen to the implicit praise of the heroic warriors defending our flag during the war of 1812. It doesn't get much more mainstream than that.

3. Cao Qin sees this as a thing of the past. He continues, "However, after the two destructive world wars of the past century, it seems that the tendency in the western world has turned to the other extreme. Although absolute pacifism has always been dismissed by most people as utopian and even counterproductive, the conventional presumption in the international arena nowadays is that we can hardly be too cautious

on the decisions to go to war" (Qin 2020, p. 164). I'm not convinced. Certainly, calls for caution are allowed to be expressed in mainstream venues, and can be dominant *in some circles*. And it is fair to say that there has been progress. But the drums of war often still beat loudest in the U.S. mainstream press. For instance, the glorification of U.S. military power and destructive technology took center stage during the "shock and awe" period of the first Gulf War. Though killing safely from a distance is in some sense cowardly, it seemed to offer a vicarious experience of masculine rage and catharsis.

4. In the Chinese tradition the military commander has not only military expertise but also moral character. He is characterized by the superior character of an exemplary person (*junzi*) (Ames 1993, p. 87). In the Confucian tradition, "Exemplary persons (*junzi*) are not implements" (*Analects* 2.12). This may be interpreted to mean that exemplary persons are not to be used as instruments or, as D.C. Lau explains it, that they are not specialists who are "designed for a specific purpose only" (Lau 1979, p. 64). The Chinese assumption is that character has an effect on how successful you are in general, and in warfare. Sunzi states: "Command is a matter of wisdom, integrity, humanity, courage, and discipline" (Ames 1993, p. 103).

5. Noddings cites a number of similar points made by significant thinkers. For example, Paul Tillich notes, "the Greek word for courage is synonymous with manliness" (Noddings 2006, p. 38). William James considers whether war will "be our only bulwark against effeminacy . . ." (p. 39). And Virginia Wolf characterizes war (scornfully) as "manliness [which] breeds womanliness" (p. 40). Noddings herself also considers the corollary to men's obsession with being courageous in order to fulfill their manliness: "To be a real woman, must one encourage, admire, and perhaps even envy 'real' men?" (p. 40). These various authors, Noddings, Tillich, James, and Wolf, are all at least uncomfortable with the association of manliness and participation in war. But the point is that they all recognize this association.

6. Ruskin seems here to be articulating a prevailing view in order to set up his critique, which immediately follows: "A state without virtue, without laws, and without honour, he is bound *not* to defend; nay, bound to redress by his own right hand that which he sees to be base in her."

7. While there is an element of this in early Confucian thinking, Daoism takes a different view. According to Hans-Georg Moeller, "The world of war in this text [the *Laozi*] is not divided into 'good guys' and 'bad guys,' but in winners and losers. In fact, there is no 'just cause,' no 'good reason' to go to war in the *Laozi*" (Moeller 2006, p. 84).

8. Making a comparison with the Western model of heroism, Moeller continues: "[T]he military leader in the *Laozi* is clearly an anti-hero, he lacks all their glorious qualities. From the perspective of the *Laozi*, however, this lack is, of course, the very condition of his success—and seen from this angle, the Western hero appears more like a fool, a person who is doomed to end tragically by becoming a victim of his own activity and strength" (Moeller 2006, pp. 83-84).

9. Confucius's comment in *Analects* 7.11 seems to be more a chastisement of his reckless follower Zilu, whose question he is here answering, then a statement

regarding his philosophy regarding war. It might nevertheless be thought to reflect something of his general attitude.

10. Brook Ziporyn's translation uses the literal meaning of *shi*: "Not doing, not being a corpse presiding over your good name" (Ziporyn 2009, p. 53).

11. The advantage of maintaining a low profile is a persistent theme in the *Daodejing*. As an example, Passage 17 says that it is best for a ruler to be virtually unknown, such that his accomplishments are regarded as having happened naturally. This theme is allied with the notions of *wuwei* (non-confrontational action) and of reversion (high and low positions tend to reverse) as well as with the general strategy of taking the low position, and with acting subtly with uncommon modesty and discretion. It is exemplified in Passage 7, which says that the sage comes first by putting himself last. See also *Daodejing* 2, 8, 10, 37, 72, 77, and 78.

12. Similarly, Lo comments, "It is noteworthy that though *The Art of War* is a manual for military success, it does not glorify war, nor does it single-mindedly advocate a maximal use of brute military force" (Lo 2015c, p. 74).

13. Mozi lists the harms of war: "Those above would have no spare time to tend to the affairs of government; the officials would have no time to order their administration; farmers would have no time to sow and reap; women would have no time to spin and weave. In this way the families of the state would lose their soldiers, and the ordinary people would exchange their duties [for inappropriate tasks]. Moreover, the chariots and horses would become worn out. Of the tents and hangings, army supplies, and shields and weapons, if only one-fifth were retained, this would still be a bounty. Besides, they would also get lost along the road; and the road would be so long that provisions would not last; when the food and drink does not last, the laborers would be starving, cold, frozen, sick, and would writhe dying in the ditches along the roadside. There could be no victory. This is of no benefit to the people, but a great harm to the entire world." (*Mozi* chapter 19, "Against Aggression;" Steve Coutinho's translation from Hagen and Coutinho 2018, p. 267.)

14. Further, as Van Els notes, "Mozi makes it perfectly clear that whoever indulges in attack and annexation will never be revered as a sage-king" (Van Els 2013, p. 86).

15. As James A. Stroble notes, Mencius implicitly rejects the war-as-a-duel model for warfare when he explicitly declares, "It is not for peers to punish one another by war" (*Mencius* 7B2; cited in Stroble 1998, p. 173, using D.C. Lau's translation).

16. Confucians and Daoists seem to be in agreement here. Indeed, Confucius seems to exemplify this comment. In the *Analects*, Confucius is said to have had nothing to say about the use of force (*Analects* 7.21), and he claims to have never studied military formations (*Analects* 15.1). Indeed, the authors of the *Analects* make a point by including the detail that, after being asked about military formations, the following day Confucius left the state.

17. Roger Ames and David Hall comment, "Coercion is anathema to human flourishing, and perhaps nothing is a more poignant symbol of coercion than the instruments of war. Far from being celebrated as the trappings of the strong and mighty, weapons should be regarded for what they really are: a most unfortunate if perhaps sometimes necessary evil. . . . This chapter sets out to demonstrate the tragic character of war. . . . War ought not to be glorified. It is always a losing proposition, and there

are no victors. Although on occasion unavoidable, it is nothing better than methodical state-sanctioned killing. Even in the event of victory, triumph on the killing fields should never be confused with the proper seasons of celebration, but instead should be treated as what it is: a state funeral properly marked by grief and mourning" (Ames and Hall 2003, pp. 125–126).

18. Moeller also comments, "The military and the weapons of the state serve to deter and prevent war, and they have to be hidden away so that they will not arouse the desire to use them. A military parade is not something a Daoist sage-ruler would approve of" (Moeller 2006, p. 79).

19. *Laozi* 31 states: "Weapons are inauspicious implements. . . . They are to be used [only] when there is no alternative."

20. These comments against deceit are particularly Confucian. In contrast, Chinese military strategists, such as Sunzi, encouraged deceit (Ames 1993, p. 104).

21. P. J. Ivanhoe notes an implication of the Confucian concern for people on both sides of a conflict: "The strong emphasis on preserving the well-being of both the people and soldiers of an enemy state and the well-being of one's own people offers a strong and respectable warrant against developing or deploying weapons of mass destruction" (Ivanhoe 2004, p. 274).

22. That they must "make their argument" rather than simply forcing people to become soldiers may be related to the fact that, early in their history of warfare, offensive weaponry was superior to defensive armor. Thus, it was dangerous to coerce citizens into battle since it was all too possible for a resentful peasant-soldier to simply kill his own ruler. As Joseph Needham remarks: "It has been demonstrated that in ancient times the progress of invention in offensive weapons, especially the efficient crossbow, far outstripped progress in defensive armour. There are many cases in antiquity of feudal lords being killed by commoners or peasants well armed with crossbows—a situation quite unlike the favorable position of the heavily armed knight in Western medieval society. Hence, perhaps, arose the Confucian emphasis on persuasion. . . . The Chinese peasant-farmer could not be driven into battle . . . since he would be quite capable of shooting his Prince first; but if he was persuaded . . . that it was necessary to fight . . . then he would march" (Needham 1969, p. 201).

23. On the other hand, as Richard Jackson notes, "JWT [Just War Theory] seems to agree on WWII as being the one clear example of a just war [out of around 3,200 recorded wars]. And yet, even in this case, the atomic attacks on Japan, the terror bombing of German civilians, the failure to stop the genocide of European Jews, and the preventable Bengali famine that killed 2 to 4 million people because food supplies were sent elsewhere for military reasons are facts that tend to be excluded from the theoretical calculations about justice, relative evils, or saving lives within JWT. . . . [And] in the single case that JWT agrees is an example of a just war, there is no acknowledgment or accounting in that justification for the way in which that war led directly to the Cold War, the division of Europe, nuclear proliferation, Cold War proxy wars in Africa and Asia, and many other conflicts" (Jackson 2020, p. 47).

Chapter Three

Anticipating Confucian Just War Theory

In the previous chapter, I described the negative attitude toward war expressed by various prominent schools of thought in early China, from the military classics to the philosophical schools: Confucianism, Daoism, and Mohism. The three chapters after this one will address Confucian just war ethics and the implications for what we call "humanitarian intervention." The present chapter sets the stage for that discussion by exploring the following: (1) The way Mohists, as anti-aggression activists, appealed to the "history" of justified war *as a way of discouraging war*. (2) The sense in which Confucians, if read as doing essentially the same thing, were not being simplistic or naïve. (3) The relevance of the Daoist critique of war. (4) The issue of how Confucianism would deal with the problem of "dirty hands"—having to do wrong to do good. (5) A brief treatment of some ambiguous statements by Confucius. All this prepares us to understand what the early Confucians were up to when they discussed the ethics of war, and to appreciate that highly restrictive criteria for using the military, based on what may initially seem to be a fancifully idealistic ethical-political theory, could be reasonable for them and perhaps also for us.

JUST WAR AND THE MOHIST "IMPOSSIBLE TEST"

Mozi was one of the earliest prominent critics of Confucianism. He and his followers were essentially consequentialists. They frame war as bad all around; in addition to being bad for people, including even the aggressor, it is bad for ghosts and spirits, as well as for "Heaven," which desires life, prosperity and order.[1]

Mohists do not just condemn the wrongness and wastefulness of war in passing, they emphasize this. Regarding the *Mozi*, Hui-chieh Loy comments, "Military conflict between states and noble houses is highlighted among the most significant evils that harm the world" (Loy 2015, p. 229).[2] And Chris Fraser suggests, "Opposition to war was one of their two most prominent doctrines" (Fraser 2016, p. 203), the doctrine of impartial care being the other. War, after all, may be seen as the antithesis of impartial care. Mohists frame wars of aggression as the most extreme cases of robbery and murder (*Mozi*, chapter 17). And so it makes sense that, as Loy remarks, "the most distinctive aspect of the Mohists' positive ethical-political proposal, namely their doctrine of 'impartial caring' (*jian'ai*), is formulated in part as a response to the problem caused by military conflict" (Loy 2015, p. 229).

Mohists, like Confucians, were activists. They weren't just articulating a philosophy; they were trying to have a practical impact on the world. Regarding warfare specifically, "Mohists saw themselves as not merely in the business of laying out an intellectual case against war; they were also concerned with seeking out would-be military aggressors and attempting to dissuade them from the path of war" (Loy 2015, p. 227).[3] Loy cites an account in which Mozi is said to have cited his skill at defensive tactics in a successful diplomatic effort to prevent an attack (see *Mozi*, chapter 50).

Although they vehemently opposed wars of aggression, the Mohists did sanction defensive wars, and even went out of their way to involve themselves in them. And it is possible that even "offensive campaigns" could be justified on Mohist grounds, under some circumstances, provided that they are really *defensive in intent*. However, as Fraser notes, the relevant criteria "are strict enough that they would only rarely be met" (Fraser 2016, p. 211). As for the narrower category of *punitive* war, Fraser concludes, "probably no supposedly punitive war could ever meet the Mohists' stringent justification conditions" (Fraser 2016, p. 209). This is in large part because the Mohists implicitly require multiple undeniable signs of Heavenly approval, such as the sun rising at night and blood raining for days, along with unseasonable weather and ghost activity. In this way the Mohist doctrine, "practically removes from human discussion the issue of whether a contemplated offensive military initiative is justified, since it is Heaven which is supposed to settle the matter by publicly observable signs" (Loy 2015, p. 242). As Fraser remarks, "Since they expect Heaven's mandate to be attested by numerous publicly observed miracles, the criterion of divine sanction in practice puts the justification for punitive war permanently beyond reach" (Fraser 2016, p. 209). This is, as Paul Van Els and others suggest, "best seen as an 'impossible test' that puts the bar for permissible wars unfeasibly high" (Van Els 2013, p. 86).[4]

Fraser provides the following example: Lord Wen suggests that he would be justified in attacking Zheng because Heaven had signaled that they deserve punishment by causing three years of scarcity. Mozi responds that Heaven's punishment itself is enough and that it would be perverse to pile on. From this we can see that what makes it impossible to satisfy the necessary criterion is not just that Heavenly interventions don't actually occur. For even if they did occur, they would somewhat paradoxically undermine themselves as justifications. This is because, as Fraser notes, "the very evidence that Heaven is displeased with a state—such as that the state suffers from scarcity—is at the same time evidence that Heaven itself has already punished it" (Fraser 2016, p. 210). Although this paradox need not apply to all possible Heavenly signs, it seems to apply to the ones that may most plausibly occur. All things considered, Fraser concludes, "For the Mohists, the doctrine of punitive war has mainly a rhetorical function, as an apology for the wars of the sage-kings. In practice, without unequivocal public portents signaling Heaven's mandate, no supposedly punitive war can be considered right" (Fraser 2016, p. 210).

In chapters 4 through 6, I will argue that early Confucians were doing something similar to the Mohists. That is, they appealed to fanciful accounts of the ancient sage kings, such that the use of the military by these moral exemplars can be excused, or even praised, while its use in the present is discouraged. Although the Confucian accounts of history are somewhat more plausible than the miraculous Mohist accounts, the criteria they imply are not in practice much more satisfiable, or so I will argue.

A key point here is that Mohists were activists as much as they were philosophers. Loy explains that rather than theoretical concerns, such as precisely articulating just war criteria, Mohists strove to make the world a better place by eliminating war. He writes:

[T]heir concern with the ethics of war seems mainly motivated by the goal of preventing military aggression. . . . [W]e give the Mohists' identity as social activists less than its full recognition when we neglect the rhetorical and persuasive intent expressed in the written expression of that thought, while blaming them for not articulating a fully worked-out just war theory. (Loy 2015, p. 244; cf. p. 228)

Similarly, we give Confucians less than full recognition as social activists when we read permissive versions of just war theory into their practical efforts to avert war. And so, my analysis of the Confucian view of war and humanitarian intervention (in the next few chapters) draws attention not just to what the early Confucians said, but what they were *trying to do* when they said it.

Note also that Mohists, like Confucians, advocate exemplary civil government instead of military engagements as a way to secure a grand reputation. Fraser explains:

> A reputation for righteousness and virtue is a laudable ambition, the Mohists agree, but aggressive war is a misguided means of pursuing it. To achieve this aim, a ruler should instead conduct diplomatic relations in good faith, distribute military and economic aid generously to other states, and govern morally and effectively in his own state—thus befriending other states while strengthening his own. In this way, he would bring innumerable benefits to all and have "no match in all the world." Wars of aggression are unjustified even if driven by praiseworthy motives. (Fraser 2016, pp. 207-208; quoting *Mozi*, chapter 19)

Although there are significant differences between Mohist and Confucian visions of ideal moral governance, Fraser's description above is in broad outline the same as what I have called, at the end of the previous chapter, "the Confucian solution." And note that the promise of a grand reputation is presented as a rhetorical device intended to influence rulers (cf. *Mencius* 1A5, see pp. 95–96; and 1A7, see pp. 69–70).

ON THE SUPPOSED NAÏVETÉ OF CONFUCIAN IDEALISM

Both Mencius and Xunzi suggest that a truly virtuous ruler would be essentially invincible. In reference to Xunzi, Eirik Harris explains the key dynamic this way: "[N]ot only will the virtuous ruler's own people unify around him, when he is faced with a morally inferior opponent, his opponent's subjects will recognize his superiority and work to save him from their own violent ruler" (Harris 2019, p. 55). As Harris notes, they would do so not so much for moral reasons but out of simple self-interest. Harris asks whether Xunzi's view is nevertheless "too Pollyannaish." He answers, "If the claim were that all one needs is virtue on one's side and every battle and enemy will immediately be vanquished, the answer would certainly seem to be yes" (Harris 2019, p. 59). But Harris correctly sees that Xunzi is taking the long view, and so a key issue is sustainability. Harris suggests that, in thinking about the military, Xunzi's underlying question is: "[W]hat kind of military policy should a state have if it aims to succeed and endure over the (very) long term?" The answer is: "States which operate in accordance with the *dao* will eventually win, leading to greater stability and flourishing for all people under their rule" (Harris 2019, p. 59). That is the Confucian answer (see *Mencius* 1A7, 2A3, 4A11; *Xunzi* 15.4, 15.6a, and 9.9).

Confucians were acutely concerned with the effects of actions on character. And people's character affects future actions, and ultimately the quality of the society. That is what the Confucian emphasis on ritual propriety is all about. Virtuous and ritualistically appropriate exemplary actions inspire emulation, repetition of which develops a stable character, from which stems more virtuous action. It is a virtuous cycle. But things can spin the other way too, as both Mencius and Xunzi seem aware. So, it would be sensible for them to reason roughly along the lines articulated by conflict studies scholar Richard Jackson: "[I]n employing violence to protect a group of innocent people in the present, for example, the long-term effects will be to reinforce the discourses and psychological mechanisms that encourage future resorts to violence and the entrenchment of an ongoing cycle of violence" (Jackson 2020, p. 51). Jackson quotes Laurie Calhoun's explanation:

> Even in a particular case where it appears that waging war will maximize the utility of the people immediately concerned, it might still, in reality, be better not to resort to the use of deadly force, since the very example set by doing so reinforces the tendency in others to do so, and these effects extend far into the future. (Calhoun 2002, p. 103, quoted in Jackson 2020, p. 51)

This would be the reverse of leading with virtue. Jackson also quotes Hannah Arendt's observation that "[t]he practice of violence, like all actions, changes the world, but the most probable change is to a more violent world'" (Arendt 1970, p. 80, cited in Jackson 2020, p. 51). These are all reasonable concerns to which Confucians in particular would be sensitive.

And yet many interpreters still see the Confucian solution as naïve. For example, Ping-cheung Lo cites "contemporary Confucian commentators"[5] and a "sympathetic interpreter"[6] as finding Mencius's idealism—the notion that moral leadership can make force unnecessary—to be naïve and simplistic. And Lo himself seems to agree (see Lo 2015d, pp. 262-263 and p. 274 n25 and n27). At the same time, Lo worries that Confucian just war theory may be easily abused, leading to wars that are not truly justified. I argue that taking Confucian idealism seriously is one way to prevent such abuse, and that doing so is neither naïve nor simplistic.

First some clarification is necessary. As Harris had indicated about Xunzi, if Mencian idealism is taken to mean that the virtue of a benevolent leader could *realistically be expected* to transform and unify the world, then, yes, *that* is naïve and simplistic. But if Mencius's idealistic analysis of true kingship was a way of excusing the incursions of the ancient sage-kings while discouraging the contemporary use of military force and encouraging instead the adoption of uncontroversially benevolent governmental policies, then it is not clear that it is a simplistic approach or that it exhibits naïveté. It is at least not more so than the view that lasting peace can be achieved by a decisive stroke of violence.[7]

Consider an example that Lo himself provides. "*Heqin*" refers to the Han government's policy of diplomacy and appeasement toward the nomadic peoples who frequently raided China's boarders during the Western Han Dynasty. The alternative strategy under consideration was military engagement—arguably a policy of extermination championed with just war rhetoric.[8] As Lo explains the situation, "The court officials in power were all in favor of war, whereas the Confucian scholars outside the court firmly rejected it" (Lo 2015d, p. 254). While the court officials were nominally Confucian,[9] they are better understood as legalist[10]/realists, that is, they would justify any means that served the interests of the state.

Lo explains that the genuine Confucian scholars supported "relying more on *wen* (enculturation, moral suasion) than *wu* (coercion), upholding *de* (virtue) and discarding *li* (力 military might)" (Lo 2015d, p. 261; quoting Wang 1992, p. 507). At times, explains Lo, the Confucian scholars suggest that "the use of force is completely unnecessary and that there will be peace and harmony among nations under Han's moral leadership only if the Han government makes a greater effort in virtuous statecraft" (Lo 2015d, p. 262). This is in keeping with the views of Confucius and Mencius who also maintained that, as Lo puts it: "A well-cultivated ruler has the moral power to win hearts and minds without recourse to physical force. This enables the ruler to maintain good order within his state and transform foreigners by means of his civilizing influence" (Lo 2015d, p. 261). It is not that moral suasion is thought to provide a *quick and perfect* solution. Rather, it is regarded as the better alternative from a long-term perspective: "'*Wen* can work for the long term, but *wu* cannot last'" (Lo 2015d, p. 261; quoting Wang 1992, p. 520). All this corresponds closely to the position I defend in chapters 4 through 6, based on close readings of the *Mencius* and the *Xunzi*.

However, Lo suggests that the Han scholars who support *heqin* have taken early Confucian reasoning too far. They should know, Lo suggests, that "Confucius and Mencius are not principled pacifists; they also admit that ancient sage-kings waged just wars" (Lo 2015d, p. 262). This notwithstanding, it does seem that these Confucian scholars were following Confucius and Mencius quite closely. Idealized tales of the virtuous sage-kings are held up in order to *contrast* with current situations for the purpose of discouraging war, not for offering readily satisfiable just war criteria (see pp. 59–64, below). In these contexts, Mencius encourages rulers to focus on implementing benevolent governance as the way to become a True King. And, as Lo puts it, "True Kings, because of their virtuous rule, attract other states to submit sincerely; they employ the military only defensively. (A True King par excellence would find no need to resort to force at all as he is admired by all)" (Lo 2015d, p. 262). To be a bit more precise, the esteemed sage-kings actually did use the military to punish tyrants, but they are said to have been able to do this essentially without battles. Yet it would not have been plausible to think that what the sage-kings ostensibly did could be achieved in the situations faced by the Han Confucian scholars, or the early Confucians before them. While it may be fair to say that it would be naïve to think that the unification of the world around a moral exemplar is achievable, it is not clear that ethical decisions based on a contrast with this ideal are likewise naïve.

In the case under consideration, in contrast to the hawkish pseudo-Confucian officials who were really "crypto-Legalists and proto-realist," the genuine Confucians were trying to prevent war by giving sometimes-idealistic reasons for favoring an alternative strategy. History suggests that these Confucians were advocating the wiser approach. Both strategies were tried over a long period. In the end, Lo comments, "though the policy of *heqin* did not stop border conflicts in those 66 years, it turned out to be more cost-effective in ensuring national security" (Lo 2015d, p. 276 n36).

By dismissing "Confucian idealism" as naïve and offering instead a just war interpretation of Confucianism that aligns closely with Western just war theory, contemporary "just war Confucians" run the risk of providing ammunition for nominally Confucian warmongers, *in direct opposition to the practical intent of the early Confucians they interpret.*

Lo seems to recognize the problem, but perhaps not the extent of it, and thus not the solution. He expresses concern that the People's Liberation Army "understands just cause very broadly, namely defending national core interests" (Lo 2015d, p. 267), and he worries that academics and policy makers "align this loose understanding of just war with a Confucian understanding of just war—to punish wicked princes"[11] (Lo 2015d, p. 268). Lo suggests that "the 'just war' language game is more for politics than for ethics. It provides the political advantages of silencing political dissidents and rallying national support" (Lo 2015d, p. 267). Lo also notes:

> It became commonplace in imperial China that whenever one wanted to use military force, rulers and rebels alike would claim to be raising a "righteous army." Arthur Waley aptly observed decades ago that: "the Righteous War principle became merely a moral cloak under which to cover acts of aggression. It was in fact a mechanism, familiar enough today, for bridging the gap between the amoralism of those who actually handle the affairs of a State and the inconvenient idealism of the masses" (Lo 2015d, p. 265; quoting Waley 1939, pp. 141-142).

In addition, Lo quotes John Mearsheimer's concern over loose Confucian just war theorizing:

> Of course, this justification for war is remarkably pliable. As almost every student of international politics knows, political leaders and policy-makers of all persuasions are skilled in figuring out clever ways of defining a rival country's behavior as unjust or morally depraved. Hence, with the right spinmeister, Confucian rhetoric can be used to justify aggressive as well as defensive behavior. Like liberalism in the United States, Confucianism makes it easy for Chinese leaders to speak like idealists and act like realists. (Lo 2015d, p. 268, quoting Mearsheimer 2014, pp. 405-406)

This is precisely why it is important to respond to the "just war Confucians," as I do in the next few chapters, and develop an account of the Confucian perspective on war that is less susceptible to abuse. But for this account to seem plausible to *us* we need to be able to take Confucian idealism seriously, and so I continue to provide justifications for doing so in the next two sections, the first of which involves a certain alignment between Daoist reasoning, common sense, and early Confucianism.

DAOIST CONCERNS REGARDING MILITARY INTERVENTION AND JUST WAR RHETORIC

As described in the previous chapter, the *Laozi* "voices a deep contempt of warfare" (Moeller 2006, p. 78). In particular, it refuses to treat it as a manly means for self-glorification. Further, as Ellen Zhang notes, "There is nothing in the DDJ [*Daodejing*] to imply the idea of waging a war for a noble cause" (Zhang 2015a, p. 193), and "The DDJ does not present the idea of 'righteous war,' since the very notion entails the idea that such a war is good and the DDJ seems to reject such a judgment" (Zhang 2015a, p. 193). Zhang reasons, "It follows that the DDJ would reject the employment of any kind of offensive campaign, including those today called 'humanitarian interventions,' since the concept of 'humanitarian intervention' itself is derived from just war theory" (Zhang 2015a, p. 193). The *Daodejing* seems to allow for weapons to be used only when unavoidable (*Laozi* 31). Although the concept of "unavoidable" is not further analyzed, it presumably would include situations such as a city under siege, especially if surrender would mean death. Intervention, in contrast, by its nature, is not unavoidable.

The Daoist perspective on warfare is not the same as the Confucian perspective. Hans-Georg Moeller explains, "The Daoist general does not fight a war out of the moral necessity, he does not try to impose a good political agenda on an evil opponent—he has no moral or political agenda at all. The *Laozi* does not speak of 'punitive' wars, it does not bring anybody 'to justice.' And neither is there a semantics of war as a liberating effort" (Moeller 2006, p. 84). This seems to contrast with the Confucian perspective in every respect— Confucians *do* speak of punitive wars and in this context approve of certain semi-mythical instances of using the military to impose good political agendas by replacing evil tyrants. But perhaps the differences are not as stark as they may seem. Both Daoists and Confucians were at least suspicious of war and emphasized the destructiveness of the wars that were current or had taken place in their recent history. And they sought to have a pacifying influence.

Like the Mohists, both Confucians and Daoists appear to have had a practical intent regarding war. They sought to encourage attitudes and ways of being that would reduce the amount of violence in the world. According to Zhang, the *Daodejing* "is meant to be an inquiry into a method to prevent war from happening amid a world full of selfish interests and excessive desires" (Zhang 2015a, p. 181). It encouraged, for example, a negative attitude toward weapons, calling them "inauspicious" (*Laozi* 31). According to Moeller, "The main function of weapons in the Daoist society is to deter or prevent war. In typical Daoist fashion, they function best paradoxically. They fulfill their purpose if they are unused—only thus do they lose nothing of their efficacy, stay intact, and do no harm" (Moeller 2006, p. 79).

Early Confucians were probably aware of Daoist perspectives regarding war and sensitive to their rationales. So, because the Daoist perspective is both reasonable and potentially compatible with the Confucian view, it would not be surprising if early Confucians were influenced by it, even if their framing of issues did not always emphasize this influence—though occasionally it did.

The Daoist strategy is to encourage a return to a simple and natural lifestyle, leading by appearing to follow (*Laozi* 7), winning by not contending (*Laozi* 73; cf. 66, 81), lessening desires by modeling lowness—calling oneself "the orphaned" or "the destitute" (*Laozi* 39, 42). This is the strategy of *wuwei* 無為, literally "non-action," though it is better understood as subtle, non-assertive, non-confrontational action. The *Laozi* advises:

> Act without [appearing to] act. Serve without [appearing to] serve. . . . Plan to address difficulties while they are easy. Do great works by focusing on the minute. Difficult work always starts with what is easy. Great works always start with the minute. This is why sages persist in *not* doing what is great, and thus can succeed in achieving it. (*Laozi* 63)

From this perspective, one does not address a foreign tyranny by forcefully putting a stop to it; one takes small, hardly noticeable steps to prevent it long before it becomes a problem. The *Laozi* advises, "What is tiny can easily be dispersed. Address it while it has not yet come to be. Order it before it becomes chaotic" (*Laozi* 64).

While the Daoist strategy of *wuwei* certainly differs from the Confucian strategy of moral leadership, there is a parallel. Belief in the power of a proper model to orient others is clearly discernable, for example, in Confucius's remark, "One who governs through virtuous potency (*de* 德) is like the Pole Star: it simply resides in its place and the multitude of stars circle it in deference" (*Analects* 2.1).[12] Similarly, using the term "*wuwei*" explicitly, Confucius also said, "Was not [the sage king] Shun someone who governed by *wuwei*? What did he do? With a respectful attitude, he properly faced south, and that is all" (*Analects* 15.5). As it applies to humanitarian intervention, the strategy of *wuwei* implies a faith that "proper" leadership will, in the long run, tend to increase order and mitigate unjust conditions.

It is not clear, *prima facie*, whether this is generally true, but it is not wholly implausible. In any case, my chief concern here is to explore a Daoist analysis that suggests that the opposite of *wuwei*, that is, taking dramatic action to fix a broken situation, is fraught with dangers. The *Laozi* warns, "Those who take action ruin things. Those who grasp at things lose them. This is why the sage does not act (*wuwei* 無為) and thus ruins nothing" (*Laozi* 64). If there is an important truth in this, as it applies to military involvement in foreign problems, it increases the merit of less confrontational strategies, including what I have called "the Confucian solution." So, let's consider the reasonableness of the Daoist inclination to refrain from confronting problems militarily.

Ellen Zhang enumerates three reasons that could account for Laozi's skepticism[13] regarding the use of military force: "(1) the ambiguity between protection and destruction; (2) the ambiguity between punishment and vengeance; and (3) the euphemistic tone implied in the language of 'being just and righteous' characterized sometimes as a gesture of self-glorification" (Zhang 2015b, p. 216). Although these are put forth as particularly *Daoist* concerns, they may be regarded as *reasonable* concerns to which Daoists tend to be particularly sensitive. Zhang elaborates as follows.

First, large scale military engagements, even if well-intended, may turn out to be destructive. Zhang relates this to the Daoist concept of *wuwei*, "non-action," that is, refraining from confrontational, contentious, or coercive action. She remarks, "For Daoists, the conception of *wuwei* challenges any action that is coercive, purposive, and egocentric, and perhaps nothing is more coercive, purposive, and egocentric than violent force" (Zhang 2015b, p. 216). Indeed, it is common sense that confrontational and coercive activity tends to engender resistance and backlash.

Second, "Daoists may argue that it is difficult to distinguish rectification from revenge and retaliation" (Zhang 2015b, p. 216). This heightens the likelihood that even a sincere attempt at "rectification" will turn into a spiral of violence. Zhang explains that, on the Daoist view, using the military for punishment is likely to elicit negative responses that may well degenerate into a cycle of violence that can be quite damaging even if each retaliation is considered by the perpetrator to be just (Zhang 2015b, pp. 216-217). While temporary success is possible, the Daoist view is that "any forced or coercive success is always short-lived" (Zhang 2015b, p. 216). Because of such dynamics, it is reasonable for even a non-Daoist to think that what was intended as protective or rectifying may well turn out to be destructive, or even disastrous.[14]

Zhang's third consideration involves rhetoric. She explains that Laozi was keenly aware of the potential for moral rhetoric to be used cynically, and that moral rhetoric used to justify war would tend to turn self-agrandizing, which he viewed as ominous. Further, he was sensitive to the damage that could ensue even from an ostensibly noble war. This is why the *Laozi* cautions against any glorificatiton of war, even in victory (Zhang 2015b, p. 217; cf. *Laozi* 31).[15] Put simply, moral rhetoric supportive of war is dangerous. Considering this, relatively hawkish "just war Confucians" ought to be circumspect. Early Confucians did suggest that there were (once upon a time) just wars. Even if they did so in a contrastive attempt to prevent war, their rhetoric may be appropriated by someone with less aversion to the use of violence in a way that inadvertently subverts their intentions.

In any case, such Daoist reasoning, to the degree that early Confucians were likely to have appreciated its reasonableness, can help us understand the early Confucian perspective on war. In particular, it can help us understand why they might be reluctant to support war, favoring instead an "idealistic," non-coercive, long-term approach, even in cases for which there is a humanitarian argument for a military solution. In the next section, I continue to argue that properly nuanced Confucian idealism is not naïve, suggesting that seemingly stubborn reluctance to violate norms, even when good consequences seem thereby achievable, can make sense from an early Confucian perspective, and perhaps from ours as well.

NORMS AND CONSEQUENCES IN CONFUCIAN ETHICS

Although in much of my writing I have emphasized the flexibility Confucianism allows (see Hagen 2007, 2019, 2010b), my argument here emphasizes limits to Confucian flexibility. Even if transcendent absolutes are foreign to Confucian thought, and even though consequentialist considerations are salient in Confucian deliberations, Confucians can nevertheless be uncompromising in their refusal to deviate from norms of propriety, even when such deviations "promise" beneficial results. For if one's means are inappropriate, one's ends, however noble, are not likely to be achieved.

On the one hand, Confucianism has a distinctly conservative side. Deep respect for tradition is, as Joseph Chan has commented, "a salient mark of Confucianism" (Chan 2008, p. 114). This side of Confucianism emphasizes *li* 禮, ritualized norms of propriety. Not to be flouted, traditional norms of propriety are geared to facilitate harmonious social interaction—quite the opposite of violence. Emphasizing ritual propriety (*li*), leaders focus on being models of proper conduct and rely on their resulting potency of character (*de* 德) to inspire and transform others. In response to the suggestion that those who do not follow the way be killed, Confucius says (as I've mentioned before), "If you are after [genuine] governance, what need is there in killing? Simply desire excellence yourself, and the common people will likewise. An exemplary person's virtue (*de* 德) is the wind; a petty person's virtue is the grass. When the wind flows over grass, it is sure to bend" (*Analects* 12.19). But when it comes to dealing with foreign tyrants some may find this strategy naïve and ineffectual in relieving oppressed peoples. The only way to deal with tyrants is with force, it might be thought. But the dismal results of many ostensibly well-intended military campaigns, and the surprising successes of the non-violent campaigns of Gandhi, Martin Luther King Jr., and others (see Ackerman and DuVall 2000, and Chenoweth and Stephan 2011), suggest that dismissal of highly principled non-violent strategies on the grounds that they are naïve is inappropriate.[16]

On the other hand, Confucianism also has a potentially progressive side that is equally fundamental. The concept *ren* 仁 can be understood as doing one's utmost (*zhong* 忠) in the service of others while applying creative introspection to attempt to take on the perspective of the other (*shu* 恕, empathetic consideration). Such moral reasoning, since it stems from personal introspection rather than accumulated collective wisdom, could lead one to strategies that are at odds with traditional norms. As Joseph Chan puts it, "In each particular case, a Confucian would have to weigh the possible consequences and consider all relevant reasons in that case before he makes up his mind . . ." (Chan 2008, p. 126). One presumably does this weighing by means of a sense of appropriateness (*yi* 義), based on empathetic consideration (*shu*), in an effort to exercise *ren* (genuine humanity, benevolence).

However, if norms are worthy of being Confucian norms, they have consideration of appropriateness and benevolence built into them. They are justified by their ability to foster conditions that engender moral development and social harmony. When such norms are violated, the benefits that they support are undermined. And yet, some interpreters believe that the early Confucians accepted that there are times when even proper norms should be violated, specifically, when the situational demands are sufficiently weighty relative to the importance of following the norm in that instance. *Mencius* 4A17 has often been read in this way. In this passage, Mencius says that rescuing a drowning sister-in-law with one's hand, though ordinarily one should not be touching her, is a matter of "weighing." The problem, from this way of looking at it, is determining whether or not the conditions are weighty enough to justify violating norms. This problem becomes particularly acute when considering the justification for war. Those who advocate "humanitarian intervention," motivated by *ren*, are willing, if the need is sufficiently pressing, to transgress norms against violence. Ordinarily, acceptable initiation of violence is limited to the punishment of those guilty of sufficiently weighty crimes. While tyrants can fall into that category, war (in real cases) does not limit its violence to only those individuals.

Let us consider the passage in which Mencius speaks of "weighing" considerations. Mencius's interlocutor suggests that Mencius is being inconsistent by, on the one hand (apparently), being willing to violate ritual norms to save a drowning sister-in-law, but, on the other, refusing to violate norms necessary to save the world. Mencius replies, "To fail to help one's drowning sister-in-law is to be a beast. That men and women do not touch hands *when giving and receiving* is a matter of ritual propriety. When one's sister-in-law is drowning, helping her with one's hand is a matter of *quan* 權 (exigency)" (*Mencius* 4A17, emphasis added). The character *quan*, which literally means "weighing," is sometimes interpreted as "discretion" (as in Van Norden 2008 and Lau 1970) and understood as suggesting "there are standard ethical requirements that can be suspended in exigent circumstances" (Van Norden 2008, p. 97). This might give the impression that Mencius would judge the permissibility of war by weighing the deviation from normally applicable standards of proper conduct against considerations of *ren* (benevolence), concern for an oppressed people. If the oppression were sufficiently severe, so this reasoning goes, he would support humanitarian intervention. Something like this seems to be assumed, for example, in Daniel Bell's interpretation (which is discussed at length in the next chapter), for he reads Mencius here as arguing "that people can transgress traditional norms in hard-luck situations" (Bell 2008, p. 230).

But there are problems with this view. For one thing, Mencius does not seem to accept the premise that saving a drowning sister-in-law with one's hand violates ritual propriety. Instead, he makes a distinction between situations involving giving and receiving, on the one hand, and those involving a drowning person, on the other. So, Mencius's endorsement of using a hand to save a drowning sister-in-law, a situation in which norms governing giving and receiving do not apply, does not imply that "standard ethical requirements ... can be suspended" or that one may "transgress traditional norms in hard-luck situations." After all, as D.C. Lau points out, "the use of the hand is in itself morally neutral" (Lau 1970, p. 246). It is not as though saving a drowning person involves harming someone else, as war does.

In addition, in the passage in question, Mencius is urged to do something to save a "drowning world." The context here suggests that this would involve abandoning the restrictions implied by ritual propriety and appropriateness. As Brian Van Norden puts it, "[Mencius's interlocutor] is suggesting that, in the current era, one must abandon the prohibitions demanded by Confucian ritual and righteousness" (Van Norden 2008, p. 97). Mencius's interlocutor is implying that one must, in effect, get one's hands dirty. But Mencius rejects this suggestion. The way to save the world, Mencius replies, is to follow *dao*. He says, "When the whole world is drowning, one helps it with *dao*. When a sister-in-law is drowning, one helps her with a hand. Do you want me to help the world with my hand?" (*Mencius* 4A17). Indeed, to depart from ritual norms in the attempt to save the world makes little sense from a Confucian perspective in which self-cultivation through ritual propriety and modeling exemplary behavior are central to the achievement of social harmony. As Judith Berling points out, "The Confucian approach to peacebuilding ... calls for the transformation and harmonization of human relationships through the moral modeling of leaders and the education and moral self-cultivation of persons" (Berling 2004, p. 106). To forsake that process in order to gain some perceived advantage is to abandon Confucianism altogether.

D.C. Lau explains why the incongruity between means and ends makes departing from the way a non-starter for Confucians. According to Lau, the issue in the passage in question is "whether one is justified in using any means that involves a breach of the ethical code in order to realize the end" (Lau 1970, p. 244). According to his analysis, the reason this passage suggests that one is *not* justified in such an ethical breach is that, unlike saving a drowning person with one's hand, in which the hand is merely an instrumental means, the way is a "constitutive means" for saving the empire. That is, the way "becomes part of the end it helps realize, and the end endures so long as the means remains a part of it" (p. 245). One cannot usher in the way by departing from it.[17] Similarly, one cannot create a peaceful world by engaging in violence.

Let's look at it from another perspective. According to Sungmoon Kim's account, despite demonstrating "remarkable moral flexibility in the service of the Way," sages did *not* dirty their hands (Kim 2016, p. 167). Kim suggests that, although sages took *controversial* actions, no *actual wrongdoing* was done, nor did they compromise their moral integrity. What is important to notice is that "discretion" does not extend to abandoning the way itself. Kim provides the example of the sage-king Shun having married without first informing his parents, violating a normal expectation. Shun is excused because the worst thing one can do as a son is to have no heir, which would be the likely outcome if Shun had waited for parental permission, which was not going to be forthcoming. So, although unorthodox, Shun's action was actually an *expression of* filial piety, not a violation of it (see *Mencius* 5A2; Kim 2016, p. 160). In contrast, to engage in full-scale war, even if motivated by *ren* (benevolence), necessarily involves wrongdoing—the killing of lots of innocent people. War is not merely an unconventional manner of fulfilling the way, it is giving up on it. To resort to that level of violence is to have lost faith in the transformative power of virtue. This may be part of the reason the early Confucian telling of the most relevant histories were sanitized.

In any case, regardless of whether or not Mencius is accepting that saving a drowning sister-in-law constitutes a technical breach of ritual propriety, the following provides a reasonable way of understanding Mencius's insistence on sticking to norms. At least regarding complex matters, one should follow *dao*, a long-term strategic approach to achieving harmony, rather than resorting to measures that *seem* expedient. Saving a drowning woman with one's hand is simple, direct, and achieves its clearly desirable result immediately. In such a case, there is no issue of complex dynamics that work in indirect ways to undermine the intent of the activity. But in complex matters, such as engaging in warfare, the risk of subtle dynamics and complex feedback chains undermining one's intentions are considerable, as Daoists emphasized. In such cases, one ought to stick with the general strategy.

The notion that the uncertainties of complex dynamics play into Confucian conservatism is somewhat speculative, but reasonable. It is clear that Xunzi advocates sticking with reliable strategies rather than abandoning them in favor of seemingly clever schemes. And we know that Mencius reflected on the relevant types of dynamics. For example, he says, "If you kill someone's father, someone will likewise kill your father. If you kill someone's brother, someone will likewise kill your brother" (*Mencius* 7B7).[18] And also:

> Those who attain the way have much support. Those who lose the way have little support. At the extreme of little support, even close relatives turn against them. At the extreme of much support, the whole world aligns with them. [Suppose] someone with whom the whole world aligns attacks someone who is betrayed [even] by close relatives. Thus, exemplary persons either do not wage war, or when they do victory is certain. (*Mencius* 2B1)[19]

Consider also the following: With regard to people who are similar to Confucius, Mencius says, "If they could obtain the whole world by one dishonorable (*bu yi* 不義) act, or by killing one innocent person, they would not do it" (*Mencius* 2A2). Van Norden comments, "Mengzi [Mencius] (and Zhu Xi) place great emphasis on using 'discretion' (4A17) to flexibly respond to complex circumstances. However, this verse [*Mencius* 2A2] makes clear that there are some absolute prohibitions that one may not violate" (Van Norden 2008, p. 43). Similarly, Xinzhong Yao comments, "[B]oth Mengzi and Xunzi condemned the killing of an innocent man, even [if] by doing this the whole empire would be secured. This has clearly pointed to a deontological position that there is an intrinsic value in human life and that violating the principle of righteousness a war with 'good' consequences would become not good or unjust" (Yao 2004, p. 99).

In addition, there are several other passages that seem to imply that there are some things that just should not be done (including *Mencius* 7B31, 4B8 and 7A17). And in 3B1 Mencius is encouraged to "bend a foot so as to straighten three yards," namely, to make a compromise with ritual appropriateness in meeting a ruler who did not approach him properly, given that much could be gained by such a meeting. But Mencius refuses saying, "Never have those who bent themselves been able to straighten others." Mencius is suggesting that seeing a ruler in a way that violated ritual propriety would not, in the end, actually be constructive.

Making a similar point, a follower of Confucius, Master You, suggests that creative efforts to produce harmony that breach traditional norms *don't actually work*. He says, "Of the uses of *li* (ritual propriety), producing harmony is the most valuable. . . . But when things are not working . . . [and] one attempts to harmonize without guiding one's efforts using *li*, this also will not work" (*Analects* 1.12). Similarly, Xunzi stresses the fundamental importance of sticking with ritual norms. He writes, "Everywhere in the world, those who follow [norms of ritual propriety] are orderly; those who do not are in chaos. Those who follow them are secure; those who do not are in danger. Those who follow them live; those who do not perish" (*Xunzi* 19/92/9; 19.2c).

Such comments provide general reasons to suspect that Confucians would not condone breeching serious norms in the service of their larger project, or in pursuit of other goods. For even well-motivated breeches are not likely to *actually* produce the desired result.

In an essay titled, "Should I Help the Empire with My Hand?" David Mathies appeals to similar reasoning to conclude:

> The Confucian, and specifically Mencian, tradition offers an alternative paradigm of power which dissolves for me the conceptual discontinuity between means and ends. Just as Mengzi [Mencius] denies that the Way can be realized

in the world by living out any lesser version of it, a more peaceable and harmonious world cannot be realized through violence. It is thus my suggestion that in response to the clean hands critique, we may well echo Mengzi and ask how can we bring peace to the world with blood on our hands? (Mathies 2011)

Put more briefly, the point is this: "If you attempt to use violent means to achieve a peaceful end . . . you will fail."[20]

Some problematic situations have no easy fix. The best one can do is to focus on being a model of proper conduct oneself and hope for the best. As Mencius says: "Exemplary persons conduct themselves according to a proper model (*fa* 法) and simply leave the rest to the forces of circumstance" (*Mencius* 7B33). He also says, "Cultivate your self and accept what comes. This is the way to take a stand amid the forces of circumstances" (*Mencius* 7A1). In the context of justifying war, arguably, this is not being naïve; it is refusing to be overly optimistic about what may be achieved through violence.

CONFUCIUS MAY NOT HAVE BEEN PERFECT

In the following chapters, I primarily interpret the views of Mencius and Xunzi, who discuss war far more than does Confucius. Confucius's views regarding the legitimacy of war are far from clear. Nevertheless, a few of Confucius's remarks about war found in the *Analects* are worth mentioning.

In *Analects* 16.2, Confucius says, "When *dao* prevails in the world . . . punitive expeditions issue from the emperor," rather than from the feudal lords or from ministers. If the *dao* had prevailed, the emperor would have had the capacity, presumably, to punish misbehaving lord-protectors with little difficulty. Confucius seems here to suggest that he, like Mencius (7B2), did not approve of peers attacking each other, which we can imagine would be a messier, potentially much more destructive situation.

However, Confucius may not have been entirely consistent on this point. After Chen Chengzi of Qi assassinated his ruler, Duke Jian, and usurped the throne, Confucius reported it to Duke Ai of Lu and requested that Duke Ai "punish him" (*Analects* 14.21). The *Zuo Commentary* provides an account of the event in which Duke Ai suggests that Lu is too weak to accomplish this. To this Confucius is said to have argued: "Chen Heng has assassinated his ruler and half of the people of Qi are disaffected. With the Lu forces augmenting this half of the Qi population, Lu can take the victory" (Slingerland 2003, p. 163; his translation). This account, in my own view, makes Confucius appear to lack due consideration for the lives that would surely be lost in pursuit of ensuring the just punishment of a single individual, as though

Confucius were putting a questionable conception of *yi* above *ren*. But much remains obscure to us about the situation.

On another occasion, Confucius was summoned by the leader of a revolt, and Confucius was inclined to meet with him. Confucius's follower, Zilu, expressed disapproval. In response, Confucius said, "If there is someone who would employ me, could I not establish a Zhou of the East?" (*Analects* 17.5), presumably referring to his hope of "re-establishing the Way of the Zhou with Lu as its political center" (Slingerland 2003, p. 202). Confucius seems here to be willing to support a revolt, as well as appearing to be, frankly, a bit of a narcissistic opportunist. But it is not entirely clear what Confucius was planning or thinking. In any case, the violence was likely to transpire regardless of whether Confucius involved himself. It is not clear that he was endorsing violence, and he may even have been intending to try to reduce it.

In addition, there is a sequence of passages in the *Analects* in which Confucius suggests that the people must have been instructed by a good person for seven years before they could be trusted with weapons (*Analects* 13.29) and that to send them to war without instruction is to "throw them away" (*Analects* 13.30). Though the kind of instruction in question is unclear, it is generally thought that it would be "primarily focusing on moral education" (Slingerland 2003, p 152). While these passages are enigmatic, one practical implication of Confucius's statements seems to be that any contemporaneous military plans should be delayed, probably for at least seven years, perhaps indefinitely.

A similar passage in *Mencius* might help us interpret *Analects* 13.30. Mencius said, "Employing the people [in battle] without instructing them is called 'destroying the people'" (*Mencius* 6B8). Though there is more context here, the precise import of this statement is still not entirely clear. What is clear, however, is the *function* that the statement is serving in this case, namely, to discourage a potential war. Mencius states his opinion regarding the situation in question explicitly: "Even if [the state of Lu] triumphs over Qi and acquires Nanyang in a single battle, it would still be unacceptable." He then provides a complex explanation based on the proper size of Lu and the fact that taking from one to give to another is not something that a person of *ren* would do. He concludes by expressing repugnance toward the idea of achieving an inappropriate end, *most especially in a way that involved killing people*, and then he says, "The exemplary person's service to his ruler consists entirely in diligently guiding him to accord with the way (*dao*), aspiring to benevolence." Mencius redirects his interlocutor from war to proper, benevolent governance, as he does repeatedly (perhaps most clearly in *Mencius* 1A7).

I mention these instances from the *Analects* for the sake of completeness, and I acknowledge that they present a problem for my claim that there is a relatively consistent early Confucian position on the issue. But these passages do not clearly undermine my thesis. At worst, Mencius and Xunzi were more thorough and consistent than Confucius in applying Confucius's ideals, such as the one expressed in the rhetorical question: When the virtue of leaders is like the wind blowing over the grass, "what need is there in killing?" (*Analects* 12.19). In any case, I can be satisfied with claiming there is a consistent Mencian-Xunzian view on warfare which, for convenience, I will continue to refer to as "Confucian."

A number of scholars have recently supported the idea that Mencius and Xunzi have a relatively permissive position with respect to the use of the military. Their arguments assume that humanitarian concerns can justify violence, and lead to the support of so-called "humanitarian intervention." Note that the common justification for humanitarian military intervention parallels the justification of torture in a ticking-time-bomb scenario.[21] In such a situation, justification for torture, which is *prima-facie* wrong (or inhumane, *buren* 不仁), is based on larger humanitarian concerns—concerns for the welfare of others. It is, in other words, justified by altruism, benevolence, or, in Confucian terminology, *ren*. This is precisely the justification for humanitarian intervention. And yet, though direct textual evidence is lacking, it is hard to imagine that Mencius would condone torture even to achieve some good end. After all, as we saw above, he is not even willing to meet a ruler who approached him improperly (*Mencius* 3B1), nor is he willing to deviate from certain norms "to save a drowning world" (*Mencius* 4A17). Arguably, asking "Who would Mencius torture?" is kind of like asking "Who would Jesus bomb?" If it is hard to imagine that Mencius would condone "morally motivated" torture, those who suggest he would nevertheless support other forms of violence to produce good results have the burden of providing textual proof. After all, Mencius said, "To kill a single blameless person is not to be *ren*" (*Mencius* 7A33). And yet innocent people are killed in large numbers in any genuine war. And, according to Mencius, killing indirectly with misrule is just as bad as killing directly with a knife (*Mencius* 1A4). So, the fact that the ruler or proponent of war does not directly engage in killing will not get him or her off the hook. This would seem to make genuine war, even if motivated by laudable ends, at least tricky for Mencius to justify.

Over the next three chapters, I address the specific textual arguments regarding the supposed support of war and humanitarian intervention found in the texts *Mencius* and *Xunzi*. I will show that the Confucian solution is not war but rather efforts to help others in ways consistent with proper conduct. Chapter 4 focuses extensively, though not exclusively, on Daniel Bell, who does acknowledge that, compared to traditional Western theories of just and unjust wars, "Mencius, and the ancient Chinese texts more generally, have been less willing to embrace what moderns would consider to be evil in the name of doing good" (Bell 2008, p. 239). But Bell's concession does not go nearly far enough. For, as I have argued here, while honest supporters of humanitarian intervention may be motivated by *ren* (benevolence) to rescue others from suffering, it is reasonable for Confucians to doubt that a fiercely resisted intervention would, on the whole, actually reduce suffering and promote harmony. From a Confucian perspective, there is a better way. That better way is the "Confucian solution," the slow and steady effort to help others in ways that do not involve violence or other serious breaches of proper conduct.

NOTES

1. It may be tempting to view the Mohist understanding of *tian*, "Heaven," as godlike, with intentions purposefully enforced with punishments. While that view is not completely inaccurate, rather than a capricious divine will, *tian*'s "intentions" are "more an established, observable, and predictable set of inclinations" (Ivanhoe and Van Norden 2001, p. 90 n48); they are "implicit in the natural workings of the Cosmos" (Hagen and Countinho 2018, p. 254). Steve Coutinho observes, "The text retains a stubborn ambivalence between natural and divine interpretations of '*tian*'" (Hagen and Countinho 2018, p. 254). What is important for us here is that "*tian* desires what is right (*yi*)," which is associated with life, prosperity, and order (*Mozi*, chapter 26). War, in contrast, is associated with death, impoverishment, and chaos. See *Mozi*, chapter 19, and Fraser 2016, pp. 206-207.

2. Loy cites *Mozi* 14.2, 15.2, 16.1, 25.13, and 32.5; Mo 2010.

3. Loy reiterates this idea in the conclusion of his essay, stating that Mohists were not only driven by abstract and theoretical concerns, but also by the need to be effective in their strategic efforts to eliminate war. See Loy 2015, p. 244.

4. Van Els credits Michael Nylan for suggesting the phrase "impossible test." Loy adopts the phrase as well: "The Mohists, it would seem, have set up an 'impossible test' for justifying offensive wars" (Loy 2015, p. 242).

5. Lo cites Xu 1979, pp. 167–170.

6. Lo cites Yuan 1992, pp. 108–109.

7. One could also argue that it is no more naïve than just war theory's "naïve assumptions about violence, namely, that violence can be used as a tool or instrument by states that then remain unaffected by participating in violence" (Jackson 2020, p. 50).

8. Lo explains that the proto-realists in this debate "abuse the language of *yizhan*," a term approximating "just war," sometimes using it loosely to suggest a battle between the good and the bad (Lo 2015d, p. 265).

9. The court officials were nominal Confucians, Lo explains, in the sense that they cited Confucian texts in making their arguments. However, according to Lo, their

political philosophy was really closer to Legalism than to Confucianism (Lo 2015d, p. 274 n22).

10. Legalism refers to a school of thought in early China that emphasized rewards and strict punishments. Lo suggests that the term "legalism" is misleading and that "coercionism" would be a more accurate label (Lo 2015d, p. 272 n4).

11. As an example, Lo cites Yan 2011, p. 41. Yan's view is discussed in chapter 4.

12. Xunzi also appropriates Daoist themes. For example, he remarks, "Great skill resides in what is not done" (*Xunzi* 17/80/17; 17.3b).

13. Zhang suggests that Zhuangzi holds a similar view: "Zhuangzi, like Laozi, holds a skeptical view about the use of force even if it is used in the name of humaneness (*ren*), justice, or righteousness (*yi*)" (Zhang 2015b, p. 215).

14. Both Mencius and Xunzi also expressed concern about perverse dynamics that play out when violence is used (as discussed below). See *Mencius* 7B7, quoted in the section below. See also "Tongdong Bai: Some Points of Agreement" in chapter 4, and "Xunzi on War and Humanitarian Intervention" in chapter 5.

15. Moeller makes some related observations: "The *Laozi* does not make any rhetorical attempts to adorn warfare at all. In this text, war is primarily seen as a social disaster" (Moeller 2006, p. 84). According to the *Laozi*, war is best avoided. And if unavoidable, it should be waged so as to minimize the carnage. Moeller notes that there is "[no] need of moral glorification" in this. And even "talk of a 'just,' a 'necessary,' or a 'liberating' war can, like heroism, appear as the somewhat presumptuous and pompous self-aggrandizement of a social loser" (Moeller 2006, p. 84).

16. Erica Chenoweth and Maria Stephan argue that the relative success of nonviolent campaigns over violent ones is attributable to the fact that "nonviolent campaigns facilitate the active participation of many more people than violent campaigns" (Chenoweth and Stephan 2011, p. 10).

17. The idea that one ought not depart from the way is a consistent theme in Confucianism. For example, the *Zhong Yong* states, "*Dao* should not be departed from, even for an instant; if it could be departed from it would not be *dao*." (*Zhong Yong* 1). Mencius says, "Because [scholar-officials] do not depart from the way (*dao* 道) when they become prominent, the common people do not lose confidence in them" (*Mencius* 7A9). And Xunzi writes, "Those who depart from *dao* and make choices from personal inclinations do not understand that on which fortune and misfortune depends" (*Xunzi* 22/112/2; 22.6b). In addition, regarding norms of ritual propriety (*li*), Confucius says, "Do not look in a way which is not *li*, do not listen in a way that is not *li*, do not speak in a way that is not *li*, do not move in a way that is not *li*" (*Analects* 12.1). I would argue that there is considerable room for personal adaptation of norms of propriety within the way—that it is a *broad* way (see *Analects* 15.29). But it is not so broad as to endorse violence against the innocent, which war inevitably involves.

18. The relevance of this passage as well as Berling's analysis of Confucian peacebuilding and D.C. Lau's analysis of *Mencius* 4A17 (both mentioned above) were brought to my attention by David Kratz Mathies. For further elaboration of these themes, see Mathies's 2011 essay, "Should I Help the Empire with My Hand," the conclusion of which is quoted below.

19. Presumably drawing from this and other passages, Xinzhong Yao interprets as follows: "Possessing the Way, a ruler would be followed, supported and welcomed by the people; while departing from the Way, he would see the people abandon him, hate him and fight against him. It is from this understanding that Confucians argue that the just war is one that wins without fighting and killing" (Yao 2004, p. 98).

20. This quotation is taken from Mathies 2011, in which Mathies likens Mencius's point to an idea espoused by Gandhi. Mathies is in turn quoting Juergensmeyer 2005, p. 39.

21. Richard Jackson makes a similar point about Just War Theory: "The logic of JWT can bleed over into other areas of human interaction so that sweat shops or torture become excepted and legitimized on the basis of similar ethical reasoning to that of JWT" (Jackson 2020, p. 51). Mencius, as a master of metaphor and analogical reasoning, is not likely to have been oblivious to the relation between morally motivated war (killing people in order to achieve some laudable end) and other unsavory actions that may be taken to achieve the same. Presumably, these are among the things that, according to Mencius, a good person just will not do (see *Mencius* 7B31, 4B8 and 7A17).

Chapter Four

Mencius on War and Humanitarian Intervention

A number of prominent interpreters have suggested that early Confucians would support what we now call "humanitarian interventions," albeit with some particularly Confucian nuances. I argue that this view is at least misleading. I maintain that the key criterion for the appropriate use of offensive military force that is implied in the early Confucian texts *Mencius* and *Xunzi* is this: on account of the responsible leader's moral virtue, overwhelming support of various kinds ensures that little resistance would be encountered, and thus the carnage associated with genuine war would not be required. Since this condition is not actually encountered in real circumstances, early Confucianism can provide support for resisting calls to use offensive warfare for humanitarian purposes.

This chapter focuses on Mencius, and interpretations of Mencius's view of humanitarian intervention, especially that of Daniel A. Bell. In the next chapter I argue that Xunzi holds a view similar to the one I attribute to Mencius. This suggests that there is a relatively consistent early Confucian perspective on warfare that is, in its practical implications, more pacifistic than recent commentaries suggest. Chapter 6 then addresses the work of Sumner Twiss and Jonathan Chan, who agree that Mencius and Xunzi share similar views on the use of the military (and so address them together) but argue that these early Confucians would be supportive of humanitarian interventions in realistic circumstances. I maintain that their arguments do not succeed. In an addendum to the present chapter, I argue further that Mencius's attitude toward defensive war is more ambiguous than commonly thought. I begin, here, with an overview of competing interpretations.

AN OVERVIEW OF COMPETING INTERPRETATIONS

The ideas defended in this and the following two chapters, in their general outline, are not new. Early Confucians have often been understood as maintaining fairly pacifistic positions. For example, John Fairbank, who Ping-cheung Lo describes as "the earliest American advocate of Confucian/Chinese pacifism," explains the Confucian position this way: "The superior man . . . should be able to attain his ends without violence. This was because of the optimistic belief that virtuous and proper conduct exerted such an edifying attraction upon the beholder that he accorded moral prestige to the actor. Right conduct thus gave one moral authority, a kind of power" (Fairbank 1974, p. 7). And Xinzhong Yao has argued, "Confucianism should not be defined as a just war theory, because it has a fundamental inclination towards ethical pacifism, namely, taking moral influence rather than war as the solution of disorder and conflict" (Yao 2004, p. 89).

However, especially over the last couple decades, a number of scholars have argued for a "just war" interpretation of early Confucianism that is relatively open to the use of military violence, so long as the aims are humanitarian. War is treated as a legitimate tool for maintaining order and rescuing oppressed peoples. Indeed, it is not uncommon for statements such as the following to be presented as though they were straightforward matters of fact: "Confucians would approve the use of force by one state against another state for the protection against abusive rule in the latter if properly carried out" (Chen 1999, p. 35; cf. Lepard 2002, pp. 90–91).[1] Such claims find support in the work of Daniel A. Bell, Tongdong Bai, and Yan Xuetong, as well as the collaborative work of Sumner Twiss and Jonathan Chan. Both Bell and Bai are explicit in their view that Confucians ought to support some humanitarian interventions. Bai writes, "When it is humane and strong, a state may need to engage in a war of liberation" (Bai 2012, p. 50). And Bell explicitly argues that there are situations in which, compared to liberals, "Confucians may be *more likely* to support punitive expeditions," such as cases in which famines are deliberately created (Bell 2008, p. 242, my emphasis). Yan makes the more general claim, "Confucius, Xunzi, and Mencius . . . are not opposed to the use of violent force to maintain order" (Yan 2011, p. 39). Although not all interpreters agree that early Confucians condoned war for these types of reasons,[2] others go further than Bell, Bai, and Yan. For example, Ni Lexiong has gone as far as to claim that Mencius viewed war as a way of realizing one's full human potential. Below, I will say more about the evidence presented by Yan, Bell, Bai, and Ni, concentrating mostly on Bell's analysis. In chapter 6, I also consider the collaborative work of Twiss and Chan, who argue that Mencius and Xunzi maintained that "serious misrule" justifies military intervention, if necessary, to put things right. They write, "[B]oth of them justify the legitimate use of military force in terms of interdicting and punishing . . . aggression and tyranny" (Twiss and Chan 2015a, p. 95). They suggest that such interdiction and punishment is not only legitimate, according to Mencius and Xunzi, but morally required.

I argue that statements such as those quoted above are at least misleading. For all *practical* purposes, I maintain, early Confucians opposed offensive wars, including those supposedly waged for humanitarian reasons. Even though they found no fault in the military expeditions of the great kings of the distant (semi-mythical) past, who serve as paragons of virtue, they nonetheless counseled against actual war-making at nearly every turn. They typically did this by providing comparisons with idealized stories, perhaps because they judged that it would not be rhetorically effective to offer a rigidly dogmatic rejection of the use of force. To extract *weak versions* of the principles seemingly implicit in these stories as a way of justifying war is a perversion of the intended function of such stories.

In the end, I find that for offensive military action to be appropriate from an early Confucian perspective the intervening ruler must be a "true king" who exhibits genuine humanity (*ren* 仁)—a very high standard. And the proof that a ruler qualifies, which thus serves as a condition itself, is that, by virtue of the overwhelming support, not only of his own people but also of the people of the state to be subject to the intervention as well as the leaders of neighboring states, *no real war need actually take place*. In other words, due to the lack of resistance, the use of force amounts to little more than a *show* of force. And consequently, though the tyrant and his close circle may be killed, there would be relatively little bloodshed, especially among the innocent. But since these conditions are not met in real cases, the practical implication is that rulers are advised to cultivate their own virtue and engage in more compassionate government themselves. To be regarded as a reasonable approach, we must understand that, despite some grandiose language, this is not really expected to deliver a quick fix to foreign despotism. Rather, it is a long-term strategic approach that avoids making matters worse with ill-fated (and often dishonest) attempts to create peace with violence.

I address several interpretations of Mencius's view of just war and humanitarian intervention. I begin by briefly addressing the position of Yan Xuetong and then, in much greater detail, the treatment of Daniel Bell. I then briefly address the somewhat alarming position of Ni Lexiong. While Tongdong Bai's position is similar to Bell's, I delay my discussion of Bai in order to conclude the main body of this chapter with qualified points of agreement, with both Bai and Sungmoon Kim. Finally, in an addendum, I also argue that, although Mencius is often regarded as approving of or even advocating defensive war in some cases, a reasonable alternative interpretation exists.

YAN XUETONG ON MENCIUS AND JUST WARS

In his book *Ancient Chinese Thought, Modern Chinese Power*, Yan Xuetong, an influential Chinese intellectual who advocates combining moral leadership with military strength, writes, "Some claim that Confucius and Mencius advocate 'no war' and are opposed to all war.[3] In fact, they are not opposed to all war, only to unjust wars. They support just wars" (Yan 2011, p. 35). Yan reasons, "Mencius . . . differentiates between just and unjust wars, showing that he believes that war is an instrument designed to implement justice or injustice and that it is a tool humans use to realize their goals" (Yan 2011, p. 35). But merely making a distinction between just and unjust war does not show that one believes that there are likely to be *real* circumstances in which war may be legitimately used as a tool to realize goals. Indeed, in the passage Yan cites here, Mencius states, "In the Spring and Autumn Period there were no just wars" (Yan 2011, p. 35; *Mencius* 7B2).[4] It is at least a little odd that a passage in which Mencius asserts that there was not a single justified war in the three-hundred-year period just preceding his own time[5] is treated as proof that Mencius was relatively open to the use of war as a tool to realize one's goals.

The following is my slightly extended translation of the relevant passage:

> During the Spring and Autumn Period there were no appropriate (*yi* 義) wars (*zhan* 戰). There were only some that were better than others. Punitive-military-expeditions (*zheng* 征) involve the superior (*shang* 上) striking (*fa* 伐) the inferior (*xia* 下). Enemy states [of similar moral status] do not engage in punitive expeditions (*zheng* 征) against each other. (*Mencius* 7B2)

Here we get a hint regarding Mencius's governing criterion, namely, only the "superior" (*shang* 上) can engage in punitive expeditions. Whether *shang* here is intended to be comparative or superlative is not perfectly clear, though *Mencius* 2B8, discussed below, suggests that it is the latter. In any case, it is implausible that marginal superiority would justify war for Mencius. I will argue that it turns out that there must be such a moral differential between the tyrannical target for "punishment" and the genuinely humane king who leads the expedition against him that the people of the invaded country uniformly welcome, rather than resist, the invading force. Thus, there would be, in such fanciful conditions, very little actual violence.

DANIEL A. BELL ON MENCIUS'S CRITERIA FOR HUMANITARIAN INTERVENTION

Daniel Bell acknowledges that early Confucianism seems to prohibit the use of force in an "ideal world," namely, in "a harmonious political order without state boundaries and governed by a sage by means of virtue, without any coercive power at all" (Bell 2008, p. 227). Nevertheless, Bell writes, "[T]he passages on warfare provide direct evidence that Confucius and Mencius allowed for the possibility that the use of force can be justified in nonideal situations" (Bell 2008, p. 230). The clearest evidence seems to come from *Mencius* 7B4 and 1B11, in which Mencius condones Tang's use of "force." Tang is described as marching his armies this way and that with the only complaints coming from those who had to wait for his arrival. Note that this case occurs in a *nonideal* situation, in Bell's sense. But it is not realistic either; it is a semi-mythic and *idealized* tale. In my analysis, in effect, I replace Bell's ideal world vs. nonideal world distinction, with an idealistic vs. realistic one (though I continue to refer to Bell's distinction in my critique of his position). In contrast to Bell's position that, according to early Confucianism, war is not justifiable in an ideal world but can be justified in a nonideal one, I maintain that in early Confucianism offensive military force can be justified only in idealized circumstances, not in realistic ones.

Bell argues that Mencius supports both defensive war and humanitarian intervention. Here I address only justification for humanitarian intervention. (I address Mencius's stance on defensive war in an addendum at the end of this chapter.) Bell writes:

> The second kind of just war [in addition to defensive war] approximates the modern idea of humanitarian intervention—Mencius labels these wars "punitive expeditions" (*zheng* 征). States can legitimately invade other states if the aim is to bring about global peace and benevolent government. Certain conditions, however, must be in place" (Bell 2008, p. 235).

Bell lays out the following conditions:

> "First, the 'conquerors' must try to liberate people who are being oppressed by tyrants." (p. 235)

> "Second, the people must demonstrate, in concrete ways, the fact that they welcome their conquerors," and "this welcome must be long-lasting." (p. 236)[6]

> "Third, punitive expeditions must[7] be launched by rulers who are at least *potentially* virtuous." (p. 236, emphasis in original)

"Fourth, the leader of justified punitive expeditions must have some moral claim to have the world's support." (p. 237)

In sum, according to Bell, a "*potentially* virtuous" ruler with "*some* moral claim" to wide support may use military force to liberate people from oppressors, so long as the oppressed people demonstrably welcome this.

The first "condition" merely seems to require that humanitarian interventions actually seek to be humanitarian, to liberate oppressed people. That seems innocuous enough as a necessary condition, but it should be strengthened. The word "try" makes it all too easy for ambitious rulers to argue that they will meet, have met, or are meeting, this criterion. And, after all, in the passage Bell cites, the people of Yan thought that King Xuan *was saving them*, not merely *trying* to save them.[8] So, some degree of likelihood, or even certainty, of success ought to replace mere effort or (purported) intention. Indeed, certainty of success by virtue of one's overwhelming strategic advantage is not an unreasonable criterion given that the great early Chinese military strategist, Sunzi, also asserted it (see Ames 1993, p. 93). The difference between Mencius and Sunzi is that, for Mencius it is moral superiority that is key, not tactical advantage. A similar idea is found in the *Xunzi* as well: "Using order to strike down disorder, one need not wait for battle to know who will vanquish whom" (*Xunzi* 27/132/25; 27.71).

Regarding the second condition—concrete demonstration of welcome—I will first address the passage that serves as Bell's primary example, arguing that the passage actually supports the view that even in the nonideal world one should focus on self-cultivation and properly governing one's own state rather than exercising military might. Following that I will argue that the condition as stated by Bell is again not strong enough. In both of these discussions I argue that *overwhelming* expression of welcome can be thought of as a sign that another condition will be met, namely, that there will be little significant resistance—which connects to the overwhelming advantage necessary for victory to be assured. Indeed, in the end, if one truly meets that condition, actual war will not be necessary. This is how Mencius can stick to his principles and at the same time praise the various "marches" of Tang.

Again, Bell's second condition requires that the people welcome the intervening armies in some concrete way. Welcoming gestures are not merely signs of approval. In addition to potentially providing actual material assistance, they constitute evidence that there would be little resistance. And that is what is critical. With little resistance, although there may be a military advancement, there is no need to engage in the kind of killing and maiming that constitute war. The part of the passage quoted by Bell in this context reads as follows:

When King Wu attacked Yin, he had over three hundred war chariots and three thousand warriors. He said, "Do not be afraid. I come to bring peace, I am not the enemy of the people." And the sound of the people knocking their heads on the ground was like the toppling of the mountain. (*Mencius* 7B4, as found in Bell 2008, p. 236)

Bell omits what comes next, which is this: "To punish-militarily (*zheng* 征) means to set-straight (*zheng* 正). If each desires to set themselves straight, what need is there for war (*zhan* 戰)?" These are the final words that sum up the point of the passage. That point is *not* that one is justified in engaging in a potentially bloody war so long as the people welcome it (and other necessary conditions are met). The point is that one should focus on setting oneself straight, and by doing so actual war can be made unnecessary. Note that this is the same passage in which Mencius says that excelling in warfare is a tremendous crime, and also, "A ruler of a state who is fond of genuine-humanity (*ren* 仁) has no enemies (*di* 敵) in the world." Although Bell himself treats this passage as informing decisions in nonideal situations, Mencius here sticks to his principles. The best thing for a leader to do when others are not straight is to focus on straightening oneself. Only when one can march unopposed into another's territory is one justified in so marching. And the lack of resistance is a sign of adequate moral standing.

It is worth noting that earlier in his essay, Bell had treated Mencius's concluding rhetorical question—"What need is there for war?"—as applicable only in an ideal world (Bell 2008, p. 231). This allowed him to dismiss it as irrelevant to nonideal situations. Bell asserts, "In a world of competing states . . . it would be foolish for states to act on the assumption that wars are unnecessary" (Bell 2008, p. 231)—which tells us more about Bell's assumptions than it does about Mencius. Bell seems to want to have it both ways: he wants to dismiss the conclusion of the passage as referring to the ideal world, while using other parts of the passage, to which the conclusion relates, to justify war in the nonideal world. Yet it is hard to see how any part of *Mencius* 7B4, read in context, can fit Bell's characterization of an ideal world, which is, as mentioned above, "a harmonious political order without state boundaries and governed by a sage by means of virtue, without any coercive power at all" (Bell 2008, p. 227).[9] On Bell's reading, Mencius's final comment turns out to be a non sequitur; on mine it suggests that rulers should focus on "set[ting] *themselves* straight" rather than on military prowess.

A more general problem with the criterion that "the people must demonstrate, in concrete ways, the fact that they welcome their conquerors" is that "the people" is not sufficiently precise. At least, it should be emphasized that this is not merely a large segment of people or even a majority. Mencius's statement, "The sound of the people knocking their heads on the ground was like the toppling of the mountain," suggests an *overwhelming outpouring* of support. Bryan Van Norden usefully captures what he calls "[Mencius's] general doctrine" as follows: "the use of military force in other states will be effective only if the people of those states uniformly regard the army as liberators rather than invaders" (Van Norden 2008, p. 82). A key word here is "uniformly," which sets a high standard. Admittedly, this could not include *everybody*. The leaders and some difficult-to-define circle of subordinates would seem to be excluded—of course *they* would not regard the invaders as liberators, for they are the ones the people are being liberated from. Indeed, Mencius cites a source that makes reference to "some who did not submit" (*Mencius* 3B5).[10] Van Norden points out that on Zhu Xi's interpretation, "'There were some who did not submit' refers to those who assisted Tyrant Zhou in wrongdoing and therefore did not submit . . ." (Van Norden 2008, p. 81). But even making an exception for such people, Mencius's general doctrine, requiring uniform approval of the people of the invaded land, sets a condition for the permissible use of an invading force that is very difficult, if not impossible, to satisfy in the real world. And, *if* it were to be satisfied, again, the tyrant's position would be so untenable that he could be removed without actual war.[11]

It is common enough for invading forces to characterize their invasion as a liberation, and to claim that they will be greeted as liberators. And, sometimes there is some degree of truth in that claim. For example, in the Iraq war, a fair number of Iraqis did seem to regard the American-led forces as liberators, at least at the beginning. But by any reasonable interpretation of Mencius's "general doctrine," that invasion was not justified. Indeed, in real cases there is almost always some significant faction of "the people" who opposed the invasion, and do not consider it a liberation, especially if the intervening force represents a country with significant interests that could be interpreted as ulterior motives—which is nearly always the case.

Conditions three and four, which require that the intervening ruler be "*potentially* virtuous" and "have some moral claim to have the world's support," are both vague standards, and arguably very low ones. (Was George W. Bush *potentially* virtuous? Does a "coalition of the willing" suffice as international support?) These criteria would likely be ineffective in preventing unjustified wars. That may seem to be merely a problem with Mencius's theory, rather than with the interpretation of it. But since Mencius consistently argued for military restraint, a relatively permissive interpretation of his implicit criteria seems problematic. If the rulers that Mencius tried to hold back were armed with Bell's set of criteria, can we doubt that they would have used it to argue that self-serving interventions were justified? For example, just as today, favors to select states and allies could be traded for statements and actions that provide "some moral claim to have the world's support."[12]

In reference to condition four, Bell cites *Mencius* 1B11 (as he does for conditions one and three), the relevant line of which reads, "The whole world was in sympathy with his cause." This seems suggestive of a significantly stronger criterion than having "some moral claim" to the world's support. Further, if we combine this with the point of *Mencius* 7B4 (discussed above in reference to Bell's second condition), the sympathy of the world with one's cause should be read not as a condition that would justify war, but a sign that a genuine war (with all the bloodshed that defines war) would not be required, for all would ally with the upright ruler against the tyrant. As Mencius says elsewhere, "If a true king were to conduct punitive expeditions against them, who would treat him as an enemy?" (*Mencius* 1A5) Presumably, recognizing the untenable nature of his situation, being hated by his own subjects and despised by his neighbors, the tyrant would eventually abdicate, or risk being assassinated or easily toppled and executed.

Finally, regarding criterion three, the evidence that Bell provides to support the claim that "*potential* virtue" is a sufficient moral quality of an intervening ruler is the mere fact that Mencius engaged flawed kings in dialog. Bell writes, "[P]unitive expeditions must be launched by rulers who are at least *potentially* virtuous . . . [O]ne can assume that Mencius bothered to talk to such flawed rulers only because he believed that they contained the seeds of virtue within them . . ." (Bell 2008, p. 236, emphasis in original). But Mencius thought that *everyone* who still counted as "human" had seeds of virtue within then, and he did not suggest that it would be justifiable for these kings to engage in humanitarian interventions. Indeed, *Mencius* 6B7 implies quite the opposite. And so, Bell's notion that mere *potential* virtue is a criterion *for just military intervention* seems to be entirely without basis. (Potential virtue is rather a criterion for simply having an audience with Mencius.) Bell adds that, alternatively, Mencius may have simply considered his royal interlocutors to have been capable of some degree of moral improvement. But *that* can't be a significant requirement for just war.

To be clear, I agree with Bell that the virtue of the invading ruler is important. Indeed, it is *more* important on my analysis than on Bell's. My complaint is that Bell has offered an extremely *low* standard without any significant textual basis. And yet there is clear textual support for a significantly stronger moral requirement (one that we will find echoed by Xunzi as well). When Mencius was misunderstood by an interlocutor as having encouraged Qi to invade Yan, Mencius explicitly denies it, and explains: "I did not. Shen Tong had asked [in confidence[13]], 'May a punitive-attack (*fa* 伐) be launched against Yan?' I answered, 'It may.' [Qi] then went and attacked it. If he had asked, '*Who* may launch an attack?' then I would have answered, 'An agent of *tian* 天 (the heavens) may do so'" (*Mencius* 2B8). Not only does this passage clarify that Mencius had not meant to encourage the attack, but it also suggests a very high moral standard needs to be met by anyone who presumes to intervene militarily. And, in Mencius's view, none of the leaders of his time (to say nothing of ours!) came close to meeting this standard (see *Mencius* 6B7). This suggests that he did not think any ruler of his time was qualified to engage in punitive expeditions. And that suggests that Mencius discussed such matters with these rulers in order to prevent them from initiating military interventions, and to convince them to better behave themselves.

Now, it must be conceded that Mencius does make approving reference to something that looks like humanitarian intervention. Specifically, as alluded to above, there are passages in which Mencius describes a time when the great King Tang's armies were on the march and whichever direction he headed the tribes in the opposite direction, wishing to be brought under his benevolent government, complained, "Why does he put us last?" (*Mencius* 7B4 and 1B11; cf. *Xunzi* 9.19a) Mencius seems to condone Tang's use of the military, in his idealized form of the story. But what is Mencius's primary objective in telling the story? It is to *contrast* those conditions with the circumstances in the case he is addressing, in which, King Xuan of Qi, after seizing the territory of Yan, failed to put genuinely humane governance into practice. By drawing this contrast, Mencius is trying to persuade the king to return hostages, to refrain from further plundering, and, after some arrangements with the people of Yan, to withdraw his forces. He is not trying to suggest criteria that he thinks may plausibly be met. His practical purpose is entirely negative, to *prevent* unjust wars, and put a *stop* to wrongdoing.

But what about the passage just prior to 1B11, in which Mencius seems to articulate clear and explicit conditions for the annexing of Yan? Mencius explains:

> If the common people of Yan would be pleased to have you annex it, then annex it. An example of someone of ancient times who did this is King Wu. If the common people of Yan would not be pleased to have you annex it, then don't annex it. An example of someone of ancient times who did this is King Wen. When a state of ten thousand chariots attacks a state of ten thousand chariots and the attacking army is welcomed with baskets and kettles of food and drink, how could it be otherwise [than that they are pleased]? [On the other hand, it is natural to] flee from floods and fire. If the people [treat your arrival] as though the water is deeper and the fire is hotter, you should simply turn around and leave. (*Mencius* 1B10)

It is important to keep in mind that the question of annexation is separate from that of military intervention. For when annexation is being discussed here, the intervention had already taken place. "The people of Qi had [already] attacked Yan and defeated it [in fifty days]" (*Mencius* 1B10).[14] It was at this point that King Xuan asked Mencius if he should annex Yan, that is, keep possession of it. Considering *Mencius* 1B11 (which suggests that King Xuan was engaging in wanton slaughter, destruction, abuse, and plunder), it seems that Mencius already knew that King Xuan was not meeting these requirements. So, spelling out these criteria in 1B10 serves as an indirect way of telling King Xuan—as Mencius does more directly in 1B11—that he "should simply turn around and leave." For practical purposes, the message is again an entirely negative one. Namely, failing to live up to benevolent governance makes one's position both unjustified and untenable.

I'll add just one further consideration. Quoting from the *Book of Documents*, Mencius says, "We wait for our ruler. When he comes, we are revived" (*Mencius* 1B11). Presumably the people would only accept someone marching in from another state as their own leader if that person had the qualities of a true king. They would not uniformly accept a marginally better replacement, or a *potentially* virtuous leader. Can we imagine a modern case in which an oppressed people, having had a tyrant removed by a foreign power, and grateful for it, would want that foreign power to stay and govern them? It is unrealistic to think that the moral credibility necessary to achieve this could be obtained.

So, in Mencius's view, the intervening ruler must be exceptionally virtuous, not just potentially virtuous or marginally more virtuous than the next ruler. How does one know if one is virtuous enough? By facing so little resistance that there would be minimal bloodshed. Although Mencius makes the following remark in condemnation of wars for territorial expansion in particular, it nevertheless shows the degree to which Mencius abhors the carnage typical of war and implies that any war that involved such carnage would not be justified.

> Confucius would renounce anyone who enriched a ruler that does not put genuinely-humane (*ren* 仁) governance into practice. How much more would he renounce those who press for war? Waging war over disputed territory fills fields with the slain. Waging war for a city fills the city with the slain. This is called, "causing the earth to consume human flesh." It is a crime not recompensed by death. Thus, those who are good at war deserve the supreme punishment. (*Mencius* 4A14)

In sum, on Bell's interpretation, a ruler who (1) tries to liberate oppressed people, (2) is welcomed (to some unspecified degree) by the invaded peoples,

(3) is "*potentially* virtuous" or capable of moral improvement, and (4) has "some claim" to international support, would be justified in launching a military intervention. On my reading, these conditions do not suffice. Rather, there must be *sufficient* support and welcome, on account of outstanding virtue, to *assure success without much loss of innocent life*. And this, in effect, rules out genuine war.

Let's consider the Gulf Wars as examples to clarify the difference between Bell's interpretation and mine. Bell argues that his criteria rule out the Second Gulf War. However, given their vagueness, it is not *clear* that they do. (1) The war was rationalized, in part, as an effort to liberate the Iraqi people. (2) It was claimed that the invading forces would be welcomed, and this was partially true. (3) Many people would argue that President Bush was at least potentially virtuous, or capable of moral improvement. (4) There was "some" degree of international support. The last point is what Bell takes to be the main distinguishing feature between the two Gulf Wars. The first was sanctioned by the United Nations; the second was not. But since Bell only requires "some moral claim to the world's support," it is not at all clear what would suffice. In any case, on my interpretation, UN authorization would not be sufficient so long as genuine war was nevertheless required. Thus, on my interpretation, both Gulf Wars would be clearly unjustified.

Bell is not entirely consistent regarding his criteria. Sometimes he implies that his conditions are stricter than his initial listing of them (in Bell 2008, pp. 234-237). For example, in reference to the justifiability of the First Gulf War, Bell writes, "punitive expeditions can be carried out by a ruler with the potential *to be a 'true king' who brings about global peace*" (Bell 2008, p. 239, emphasis added). That sounds like a much stricter standard than mere potential virtue (of some unspecified degree). And yet, strangely, Bell seems to suggest that the First Gulf War may have satisfied this criterion. Surely the notion that George Bush Sr. had "the potential to be a 'true king' who brings about global peace" is fantastical. And, indeed, Bell refers to him as a "hegemon." But Bell seems to imply that this may not matter because "the United Nations lent moral legitimacy to the war," as though the UN can serve as a proxy for a potential true king. Is a UN resolution enough to cover *both* the requirement for a virtuous leader and the requirement for international support? Before one can make an informed judgment in this case, one must consider complicating factors. The UN vote was 12 to 2, with China abstaining. According to Geoff Simons, in order to achieve this result, and thus the appearance of wide international support, "Washington set about bribing and intimidating as many UN Members as seemed necessary" (Simons 1996, pp. 196; for details see pp. 197-198.) In reality, even the support within the US government was marginal. The Senate voted to support military action by a slim 52-47 margin, and that appears to have been significantly influenced by the apparently false claims of a fifteen-year-old girl who turned out to be the daughter of the Kuwaiti ambassador (Rowse 1992). Would marginal and manipulated support really be sufficient to justify war for Mencius? I don't think so, but Bell's criteria are weak enough and vague enough that cases like this become at least arguable.

NI LEXIONG'S USE OF CONFUCIANISM TO GLORIFY WAR

Ni Lexiong's position is more extreme than that of Bell or Bai (who view Confucians as only supporting offensive war for humanitarian reasons), or even Yan (who seems to have a bit more expansive notion of Confucian just war theory, justifying war for "maintaining order"). While Bell characterizes war as "an unfortunate but occasionally necessary event stemming from difficult circumstances" (Bell 2008, p. 248 n24), Ni turns war into something of positive value. Ni argues that, on Mencius's view, humanitarian interventions would count as following the way of benevolence. Indeed, he asserts that Mencius views war as a "means of practicing 'benevolent government,'" and that it is even a way of "realizing the highest ethical ideal of 'benevolence'" (Ni 2008, p. 209).[15] The passage to which Ni seems to refer is *Mencius* 7B3.[16] Looked at critically, this passage does not show that Mencius would condone morally motivated war, much less that war is a means of attaining moral greatness. The passage reads as follows:

> Mencius said, "To trust the *Book of Documents* completely would be worse than not having it at all. I only accept two or three bits of the 'Wu Cheng' chapter. A person who exhibits *ren* (genuine-humanity) has no enemies in the whole world. So, when the most humane [King Wu] sent an expedition against the most inhumane [the tyrant Zhou], how could it be that the blood flowed so that it carried along shields?" (*Mencius* 7B3)

This passage does seem to suggest approval of King Wu vanquishing the tyrant Zhou. But it is not fair to say of this passage, as Ni does, that "Mencius firmly insists that we should use the 'army of benevolence and justice' to wage the war of 'punishing/killing the tyrants'" (Ni 2008, p. 209). No general *insistence* of this kind is made. One may at most infer that Mencius would *condone* such an expedition in some circumstances. And there is a significant difference between *condoning* something and *insisting* upon it. In addition, there is a still more important issue: Mencius would condone this kind of military action, it seems, only in conditions under which there would not be too much carnage.

Indeed, all that the passage actually explicitly says is that Mencius *can't believe* that when a truly humane king attacked a terrible tyrant there was much bloodshed. But the fact that Mencius can't believe this is itself interesting. Mencius seems here to allow his ideology to influence his assessment of historical reality to the point of disbelieving a source that has, presumably, greater empirical basis than he has himself—not to mention greater plausibility from a modern perspective (making allowance for the hyperbole). That is, Mencius expresses doubt in the *Documents* not because he has contradicting empirical information, but simply because it does not fit with his theory and the standard caricatures of the principal agents—the virtuous versus the wicked. Although no historical account should be treated as gospel, Mencius's objection to the recorded history here seems extremely weak.

Contemporary Confucians would do better to respond to what is recorded in the *Documents* by recognizing that real conflicts are almost always complex, and don't resolve neatly into good and evil. Further, even in the clearest cases, there will be many that have their interests tied up with the wrong side and will fight to defend those interests. And further, people may be bribed, duped, or coerced into supporting a tyrant. Mencius, along with Xunzi, may be right in thinking that a tyrant's support will ultimately be, by virtue of his tyranny, fragile, and that that fragility will be critical to his ultimate demise (*Mencius* 1A7; 1B10; 1B11; *Xunzi* 9.5; 9.19b; 10.13; 15.6a). But history suggests that it is wrong to think that, in real cases, this demise will occur easily and without much loss of life. Understanding this, modern Confucians ought to be all the more wary of attempts to justify war on humanitarian grounds.

TONGDONG BAI: SOME POINTS OF AGREEMENT

Bai's position is similar to Bell's. As mentioned above, according to Bai's interpretation of Mencius, "When it is humane and strong, a state may need to engage in a war of liberation" (Bai 2012, p. 50). For Bai, it seems, the chief criterion is that "the people attacked have to welcome the attack," and that welcome must be long lasting (Bai 2012, p. 51). I have addressed this issue above. Suffice it to say here that if an intervening state were humane—in the strong Confucian sense—and the welcome so overwhelming that little fighting was needed, then I would agree. But these conditions do not occur. Bai seems to have a significantly weaker interpretation of these requirements, as he suggests that they are met in real circumstances.

In any case, instead of dwelling further on our differences, I would like to endorse several points Bai makes, and explain why I think they provide additional reasons to doubt that Mencius would support relatively permissive criteria for humanitarian interventions. First, Bai acknowledges, "Early Confucians urged caution against wars, including those waged for purportedly humanitarian reasons" (Bai 2012, p. 47). On this point I don't think there is much dispute. Second, he points out, "Confucian justifications for war could well offer ambitious rulers the opportunity for pretense" (Bai 2012, p. 86). This is a serious concern given his, and Bell's, relatively weak standards of justification. Confucius and Mencius often confronted rulers who were acting on pretense (most clearly in *Analects* 16.1 and *Mencius* 1B11). Spelling out weak criteria would only encourage the kind of abuses that Confucius and Mencius tried to restrain. A recent example of this is then U.S. Attorney General Eric Holder's justification of the assassination of an American citizen in Yemen by pointing out that the fifth amendment's requirement of due process does not explicitly require a "*judicial* process."[17]

Third, discussing Confucius and Mencius, Bai notes that "If people of one's own state are treated well, but people from a different state or fief remain intractable and even pose a potential threat, we should strive to improve our civilization and moral character in order to win them over" (Bai 2012, p. 48). This reasoning seems as applicable to states that mistreat their own people as it does to states that pose a potential threat. Bai concludes that the Confucian view on war is as follows: "[1] war should be our last resort: [2] the best way to disarm a threat and prevail in the world is to become a moral exemplar" (Bai 2012, p. 48). The notion that war should be a last resort is not controversial. The problem is that advocates of any particular war generally proclaim that they *are* turning to it as a last resort, that there is, in other words, *no viable alternative*.[18] What is most significant about the Confucian approach in this regard is that it *offers a concrete alternative*, namely, to use moral power to undermine pernicious regimes. This is not an idea that is totally unheard of in Western discourse, but it is not *often* heard. And more significantly, Confucianism adds a rich theoretical structure to this idea.

Mencius 1A7 provides a good example. Here Mencius addresses a king who uses his military to try to extend his territory. While the king was not acting from altruistic motives, Mencius's treatment of this issue nevertheless has relevance to humanitarian intervention. Mencius points out that King Xuan's use of the military "endangers soldiers and ministers and incurs the resentment of the feudal lords" (*Mencius* 1A7). Mencius then deduces the king's aim and provides his assessment. He says:

> You desire to expand your territory, to have the states of Qin and Chu at your court, to rule the central kingdoms, and to pacify the barbarian tribes. Seeking what you desire in this way is like climbing a tree to catch a fish. . . . It is more perilous than that. If you climb a tree to catch a fish, although you will not catch any, there will be no calamity. But by seeking what you desire in this way, you

will exhaust yourself emotionally and physically in the process, and it is sure to end in disaster." (*Mencius* 1A7)

Here it is clear that Mencius is thinking of dynamics that tend to undermine attempts to use force in a way that is resisted and resented. And he makes the point that it is not only ineffective, but ultimately self-destructive. Mencius suggests the following strategy instead:

> Why not return to what is truly fundamental [literally, the root]? If Your Majesty governs in a way that expresses genuine-humanity (*ren* 仁), the officials of the world would all want to take their place in Your Majesty's court. The farmers would all want to plow Your Majesty's fields. Merchants would all want to house their wares in Your Majesty's markets. Travelers would all want to go by way of Your Majesty's roads. And anyone under the heavens that bore ill will toward their ruler would want to share their complaints with you. Under these conditions, who could prevent [you from achieving your ends]?" (*Mencius* 1A7)

The advantages of returning to the fundamental, namely benevolent government, seems applicable to cases of humanitarian intervention as well, for the effects mentioned would serve to increase one's influence beyond one's own state. This is not a strategy that is limited to the ideal world, it is the "fundamental" Confucian strategy.

Referencing the above passage (*Mencius* 1A7; and 1A5 as well), Bai suggests that "[Mencius's] claims to the power of humanity [*ren* 仁] are meant to inspire, and should not be taken at face value" (Bai 2012, p. 49).[19] I agree that Mencius is attempting to inspire King Xuan (as well as King Hui) to extend his compassion and thereby become a better and more effective ruler. And I recognize that Mencius, like other early Confucians, was prone to hyperbole and exaggeration. But that does not mean that he was not serious about the dynamics he describes. Admittedly, benevolent yet peaceful tactics are not likely to bring an end to distant evils easily or quickly. Yet given the serious and sure evils of war, along with the uncertainty of war's benefits, it seems that taking the longer perspective, and adopting a moral-strategic approach, rather than resorting to a violent gamble, ought to be given due consideration. Mencius provides a framework for such considerations, and, as we shall see in chapter 5, Xunzi corroborates and contributes to it.

SUNGMOON KIM ON VIRTUE AND DEMOCRATIZING MENCIUS

The position I have been arguing for shares some common ground with that of Sungmoon Kim, at least with his democratized reconstruction of

Mencius's view. Kim argues that, for Mencius's theory to be relevant to contemporary humanitarian intervention, it must be democratized such that military intervention is morally justified from the perspective of the people of the intervening state, the people of the intervened state, and also to the international community (Kim 2017, p. 187, 207).[20]

However, Kim suggests that this requires a shift away from focus on the virtue of the intervening authority. He writes, "a modern Confucian theory of humanitarian intervention, its Mencian inspiration notwithstanding, needs to be decoupled from Mencian virtue ethics and politics, preoccupied with the agent's moral character" (Kim 2017, p. 199). I'm not so sure about that. But what is important to note here is this: though Kim worries that emphasis on personal virtue cannot provide an effective check against the inappropriate use of the military, he agrees that *in Mencius's view*, the legitimate use of military force is contingent upon the intervening agent's "immaculate" moral character (Kim 2017, p. 192).

According to Kim's analysis, the only significance of the ruler's virtue is that it assures that his judgement will be good. Kim writes: "In Mencius's view, what motivates and justifies the ruler's defensive (and offensive) military engagement in the end is solely his moral judgment, which in theory is objectively good as long as he is virtuous" (Kim 2017, p. 199).[21] This description encourages the following line of thought. If a ruler is sufficiently virtuous, his *judgement* is going to be right, and so: if he judges a military intervention to be justified, then it is. That provides little practical guidance, as Kim recognizes, since judging moral character in the abstract is difficult. Understandably, Kim worries that this leaves Mencius's theory "helplessly vague," making it "impossibly difficult to constrain the ruler's decision or action in a politically principled way" (Kim 2017, p. 203).

While Kim worries that tying justification of military intervention to the character of the intervening authority ends up being too permissive, I have argued, on the contrary, it is very restrictive, *because it is so clear that nobody now qualifies*, just as nobody did in Mencius's time either. If one holds out hope that someone may qualify, the problem of determining an agent's character is mitigated by treating the salient implication of adequate moral character as the criterion: such overwhelming support that genuine war is unnecessary.

Here is where my interpretation diverges from Kim's: Mencius held the significance of virtue to be much broader than its contribution to good judgement; it has practical implications which *substantially alter the dynamics of the situation*. Without these effects, military intervention cannot be justified. And thus, these effects operate as criteria. Some of these, such as near-universal support for the intervention, Kim essentially acknowledges. He writes: "[A]ccording to Mencius's moral theory, a virtuous ruler, by virtue of his moral charisma, garners support from all around the world (*tianxia*), including the people of the state in which he intervenes, for his punitive expedition" (Kim 2017, p. 207). And, "Without explicit and overwhelming support for the external intervention on the part of the local people, humanitarian intervention by an external force is never justified, however well-intended the intervention is and however serious the local situation is" (Kim 2017, pp. 207-208). Further, as I mentioned above, on Kim's reconstructed Mencian theory any military intervention must be overwhelmingly supported by the people of the intervening state, and by the people of the intervened state, as well as garnering international approval too (Kim 2017, pp. 187, 207). In the end, Kim and I end up nearly in the same place, except that I see these criteria as tied, in Mencius's view, to the ruler's virtue.

Kim considers a situation in which a fraudulent "humanitarian intervention" results in even worse oppression than had been present before. After the fact, one can easily judge such an intervention to be wrong, and the ruler who led it to be an impostor. However, Kim writes, "Mencius offers no principled way to prevent such instances from occurring in the first place" (Kim 2017, p. 203). I have argued that, on the contrary, Mencius does accomplish this with his difficult-to-meet *indications* of adequate virtue. And what makes the level of virtue *adequate* is that the effects that serve as indications (and thus can be used as criteria) also serve to substantially change the dynamics of the situation such that the harm associated with genuine war does not occur.

There is a sense in which Kim agrees. Describing the *ideal* Mencian intervention—and all of the interventions that Mencius condones are of this variety—Kim writes: "Ideally, the conquering ruler's brilliant moral character and the moral charisma radiating from it will attract the local people to spontaneously (*ziran* 自然) submit to him without any sign of struggle or resistance" (Kim 2017, p. 193). Kim later refers to such a situation as "spontaneous harmony," in which "the people as beneficiaries of [the intervening agent's] Heavenly virtue (or beneficence) . . . respond to the new ruler's virtue with gratitude" (Kim 2017, p. 207). He remarks, "Spontaneous harmony does not require moral justification," and also suggests that this "is hard to attain in the modern world, especially between the external intervener and the people that are intervened" (Kim 2017, p. 207). I would put these points differently: Intervention *does* require moral justification. It is just that in idealized cases, in which the conditions for spontaneous harmony are present, it *is justified*. However, unfortunately, such cases don't exist in the real world.

ADDENDUM: MENCIUS ON SELF-DEFENSE

The above analysis stands on its own. But I would like to add a related discussion of Mencius's view of defensive war. Sumner Twiss and Jonathan Chan state: "[P]resumably any ruler has the rightful authority" to engage in military action for the purpose of "self-defense against invasion or impending invasion by another aggressive state." And they assert that "Mencius is particularly clear about this justifiable cause" (Twiss and Chan 2015a, p. 104).

I take no position here on whether Mencius really did find this a justifiable cause. It is, after all, more than plausible. And Mencius didn't make any clear claim to the contrary. Nevertheless, by providing a reasonable interpretation under which Mencius does not endorse defensive war, I suggest that the truth in this matter is *less clear* than commonly supposed. While Mencius *seems* at first blush to at least sometimes condone defensive war (*Mencius* 1B13), and has been read even as *advocating* it (Bell 2008, p. 231), if we consider his comments on this matter in context, *with attention to their practical effect*, we can reasonably wonder whether Mencius truly endorses even defensive war. This discussion also reinforces the view that Mencius sticks to his fundamental teachings even in his advice regarding real situations in an imperfect world.

In *Mencius* 1B13 through 1B15, Mencius addresses a problem faced by Duke Wen of Teng, whose small state is under pressure from the powerful states of Qi and Chu. He suggests that self-defense is an option. "If you defend it alongside the common people, such that they will risk their lives and not abandon you, then there is some chance" (*Mencius* 1B13). And in 1B15 he remarks that some people say, "Generations have guarded it, it is not for me [to abandon it]. Even if it means my own death, I will not leave." He suggests that taking this attitude is a legitimate option. Note that while Mencius does not fault those who feel compelled to defend their homeland, and suggests that acting on such feelings is legitimate, he does not suggest that anyone is morally obligated to engage in defensive warfare. Indeed, he suggests that abandoning one's homeland is also an option. Importantly, Mencius's (apparent) approval contains the proviso: *if* the ruler chooses to fight, he must put his *own* life on the line.

This case (specifically *Mencius* 1B13) is the principal example Bell cites to show that Mencius had different advice for ideal and nonideal cases. Bell writes, "[H]ad [Mencius] only been concerned with the ideal world, he would have urged Duke Wen to rely exclusively on moral power to deal with larger states, in the hope that virtue would attract the good will of people outside and eventually make territorial boundaries obsolete" (Bell 2008, p. 230). However, the fact that Mencius's response is sympathetic and tailored to the situation does not mean that he had two sets of standards. The important thing is whether Mencius was trying to influence Duke Wen in a way that is consistent with his fundamental teachings.

To address this, let us consider more context. In addition to the option of self-defense, in which case Duke Wen would be required to put his own life on the line, Mencius also offers an alternative. He recounts the story of King Tai of Bin, who, facing a similar situation, renounced his title, and walked away—only to have his people follow him and resettle with him elsewhere. Note that this suggestion appeals to the fundamental Confucian assumption that the influence of moral power stimulates a positive response from the people, which helps ameliorate problems. (In fact, even Mencius's proposed manner of self-defense—stimulating loyalty by putting one's own life on the line—also relies on this.) In one version of the story, strangely (given that Mencius had seemed to suggest that self-defense was a viable choice), Mencius says that it is *not* that King Tai *chose* this course, but that "there was *no alternative*" (*Mencius* 1B14). And the passage concludes: "Success depends on *tian* 天. What can you do about your situation? Simply strive to be excellent (*shan* 善), that is all" (*Mencius* 1B14).

Now, how do we make sense of all that? Perhaps we will be best guided in our understanding not by attending to the logical implications of Mencius's words so much as by noticing the likely concrete implications. When Mencius says to the king, in effect, "You could fight to your death side by side with the common people," how is that likely to be received? Not with much enthusiasm, one would imagine.[22] Then, Mencius offers a noble way out, complete with a grand precedent. And he caps it off by relieving the king of the responsibility for the result (whether people follow him or not), for success is a matter beyond a single person's control. What one controls is one's own moral conduct. Just try to do what is right—walk away, and don't worry beyond that. On this interpretation, Mencius is sticking to his general principles, not making practical compromises for a nonideal situation.

There is one further consideration here. Duke Wen is in a particularly tight spot. What if conditions had been similar, yet more favorable? Here is a clue that Duke Wen has not done all he could in creating the conditions that would have made his situation more secure, which suggests that he is not particularly worthy. In *Mencius* 1A5, 7B3, and 7B4, Mencius remarks, "Those of genuine-humanity (*ren*) have no enemies." And, in the first of those passages, Mencius goes as far as to say that, under a king who practiced genuine-humanity—that is, one who "spared punishments, and reduced taxes, so that the fields will be deeply plowed and well cultivated"—"the people could be enjoined to fashion simple staffs with which they would beat back even the well-armed and armored soldiers of Qin and Chu" (*Mencius* 1A5). Although

Mencius probably did not mean this literally, the general implication is still relevant. Namely, security is to be found in benevolent government. The ability to defend one's territory is a natural consequence of excellence in benevolent government. And, if done well, a military defense, though possible, would not be necessary. This interpretation is reinforced by Mencius's response to a similar case. In reference to the small and potentially vulnerable state of Song, Mencius concludes, "If [the ruler of Song] put the policies of true kingship into practice, then all within the Four Seas would raise their heads, look with admiration upon it, and wish that you were their ruler. [In such a case,] although the states of Chu and Qi are large, what would there be to fear from them?" (*Mencius* 3B5). Xunzi confirms this view, writing, "If a ruler desires strength, security, peace, and joy, there is nothing better than turning back to the people. If one desires to make his subjects dedicated and to unify the people, there is nothing better than turning back to sound government" (*Xunzi* 12/59/1; 12.5; cf. 9.4).

This analysis does not prove that early Confucians would never truly support defensive war in real cases, though it does raise that possibility. At least, it suggests that, from a Confucian perspective, circumspection is in order regarding the appropriateness even of defensive war, and leaders who are concerned about security ought to concentrate on moral improvement *before* they find themselves and their state in an untenable situation.

NOTES

1. Similarly, Yang Qianru writes, "Even though Mencius says that there were no just wars in the Spring and Autumn Period, he still holds that undertaking war to preserve dying states, to ensure succession for those without an heir, to remove tyranny, and to stop slaughter is right" (Yang 2011, pp. 150–151). Yang simply asserts this without citation or further discussion. Are we to believe that none of these conditions occurred during the Spring and Autumn Period—a turbulent three hundred years? (See *Mencius* 7B2.) And further, it would be odd for Mencius to be so open to various possible wars if it is true that, as Cao Qin suggests, "virtually none of the wars that happened in his lifetime could have counted as just" (Qin 2020, p. 169).

2. David Kratz Mathies, for example, argues for a position similar to mine. His conclusion is quoted in the previous chapter (p. 48). Also, Yan Xuetong cites a book by Liang Qichao, published in 1922, as arguing that Mencius was a pacifist (Yan 2011, pp. 21-22). In addition, John Ferguson makes the following intriguing claim, which Mathies alerted me to: "[Confucius] did not challenge war; he refined it, but that refinement itself proved a challenge. For a check on unrestricted war proved to be a check on war" (Ferguson 1978, p. 64). Unfortunately, Ferguson does not elaborate.

3. Yan cites Huang 1996, p. 76.

4. Here I have used the quotation as it appears in Yan 2011, p. 35.

5. The Spring and Autumn Period was 771–476 BCE, or on other interpretations, running until 403 BCE. Mencius lived 372–289 BCE.

6. Bai also stresses both aspects of this criterion, which seems to serve as the key justification for humanitarian intervention (see Bai pp. 50–51). I leave out the "long-lasting welcome" condition in the analysis below because it has limited usefulness at the time intervention is being contemplated, which is my focus. It can be useful, however, in retrospective assessments.

7. It is not perfectly clear whether Bell means that (a) punitive expeditions "must [*only*] be launched" by potentially virtuous rulers, or, (b) so long as the other conditions are met, potentially virtuous rulers *must* (i.e., *are morally obligated to*) engage in humanitarian interventions. I take Bell to mean the former. But Sumner Twiss and Jonathan Chan explicitly maintain that, for Mencius and Xunzi, morally qualified rulers have a "duty to rectify a severely tyrannical situation" (Twiss and Chan 2015b, p. 131; see p. 93, below).

8. The passage cited reads as follows: "Now the Prince of Yan cruelly mistreated his own people and Your Majesty set out on a punitive expedition. Yan's people thought you were saving them from 'flood and fire' [i.e., from tyranny]" (*Mencius* 1B11, as quoted in Bell 2008, p. 235).

9. For convenience in judging this claim, I've included *Mencius* 7B4 in its entirety here: Mencius said, "There are some who say, 'I excel at arranging battle arrays, and at warfare (*zhan* 戰).' This is a tremendous crime. A ruler of a state who is fond of *ren* 仁 (benevolence/humanity) has no enemies (*di* 敵) in the world. When [King Tang] marched (*zheng* 征) south, the Di tribes of the north were resentful. When he marched east, the Yi tribes of the west were resentful, saying, 'Why does he put us last?' When King Wu defeated (*fa* 伐) the Yin, he did so with 300 chariots and 3,000 warriors. Yet the King announced, 'Fear not, and rest assured, the common people are not my enemy.' And the people bowed down as though they had collapsed. To punish-militarily (*zheng* 征) means to set-straight (*zheng* 正). If each desires to set themselves straight, what need is there for war (*zhan* 戰)?"

10. I use Van Norden's translation here to maintain consistency with his comment.

11. According to Xunzi, the toppling of the tyrants Jie and Zhou was as easy as "punishing solitary individuals" (*Xunzi* 15/70/15; 15.1d).

12. See Simons 1996, pp. 197–198, for examples of incentives and disincentives that may have influenced the outcome of UN Security Council Resolution 678, generally regarded as authorizing the First Gulf War.

13. Earlier in the passage, the question is said to have been asked in confidence, or as Bryan Van Norden renders it, "on his own behalf." Van Norden explains that this means "this was not an official inquiry from the king about what he should do" (Van Norden 2008, p. 56).

14. Note that, by itself, being able to win an easy victory does not imply that military action was justified. In a different context, Mencius comments, "Even if [the state of Lu] triumphs over Qi and acquires Nanyang in a single battle, it would still be unacceptable" (*Mencius* 6B8).

15. The following is the context in which Ni Lexiong makes the above-stated comments. Ni writes, "Mencius firmly insists that we should use the 'army of benevolence

and justice' to wage the war of 'punishing/killing the tyrants.' In this way, war is connected with the Confucian ethical ideal through real politics. War is not only a means of practicing 'benevolent government,' it is also a means of realizing the highest ethical ideal of 'benevolence.' So for Mencius the Confucian ideas of war changed from maintaining the social and political order to protecting people's right to life and thus practicing morality and ethics. War in Mencius's thought became the tool of morality and ethics, which is another contribution of ancient Chinese thought to world civilization" (Ni 2008, p. 209).

16. There is no clear citation regarding the passage that Ni refers to here. Presumably he is referring to *Mencius* 7B3, which he cites on the following page in conjunction with precisely the same claim made here, namely, that the use of war to save people from cruel tyranny is "the realization of the highest Confucian ethical ideal of 'benevolence'" (Ni 2008, p. 210). In addition, much of what he writes here could, roughly, be thought of as being supported by that passage.

17. This case exemplifies how even strongly restrictive criteria, made explicit, can be perversely interpreted and turned into justification for what it was intended to prohibit. Attorney General Eric Holder appealed to "applicable law of war principles" in justifying the extra-judicial assassination of American citizen Anwar al-Awlaki, a suspected terrorist (Holder 2013). At the same time, Holder asserted that the requirement of "due process" guaranteed in the fifth amendment, which seems at first glance to be both strongly restrictive and appropriately nuanced, was satisfied by an "executive branch review" because the fifth amendment does not explicitly require a "*judicial* process" (Finn and Horwitz 2012). Let us carry this just a small step further. Considering the fact that the fifth amendment includes an explicit exception for "cases arising in the land or naval forces, or in the Militia, when in actual service in time of War *or public danger*" (emphasis added), the door seems open for it to be used as *positive justification* for assassinating anyone deemed by government officials to be a "public danger." But, surely, that turns the purpose of the amendment on its head. I don't know what Mencius would say about this case, but I suspect that he would be more inclined to offer a useful analogy than to clarify necessary and sufficient conditions.

18. Robert L. Holmes makes the following relevant point: "[W]e simply do not *know* whether there is a viable practical alternative to violence, and will not and cannot know unless we are willing to make an effort, comparable to the multibillion-dollar-a-year effort currently made to produce means of destruction and train young people in their use, to explore the potential of nonviolent action" (Holmes 2013, p. 167, cited in Jackson 2020, p. 55).

19. Bai cites *Mencius* 1B13–1B15 as revealing Mencius's "realistic side" (2012, pp. 49-50). These passages address the issue of justified self-defense, and I discuss them in the addendum, which focuses on that subject. I argue that it is not entirely clear what Mencius is truly trying to motivate Duke Wen to do in those passages.

20. Kim is particularly concerned about the way in which, in the contemporary world, the people's wishes and decisions can be properly processed. He makes the following quite interesting suggestion: "[T]he terms of intervention are to be negotiated . . . between the local people [of the intervened state] . . . and the intervening

state in a way that allows the former to be able to hold the latter accountable for their conduct in war" (Kim 2017, p. 207).

21. Similarly, Kim also writes, "What justifies the virtuous ruler's military intervention in . . . a troubled state is solely his moral judgment, the objectivity of which is singularly grounded in his moral character" (Kim 2017, p. 202).

22. Mencius's expectation regarding Duke Wen's reaction may have been tied to his assessment of Duke Wen's personal courage.

Chapter Five

Xunzi on War and Humanitarian Intervention

Let us turn now to Xunzi, who provides what may seem to be the clearest general advocacy for something very much like humanitarian intervention. Here I argue that Xunzi's view is similar to that of Mencius, as I've described it. Much less attention has been paid to Xunzi's view of the appropriateness of war,[1] but if, as I argue, his views largely cohere with those of Mencius, it helps us form a fuller conception of a relatively consistent early Confucian attitude toward warfare in general, and toward humanitarian intervention in particular.

YI-MING YU ON XUNZI AND WARFARE

Before delving into passages from the *Xunzi*, let's review a recent treatment of Xunzi's perspective on war by Yi-Ming Yu, a political scientist at the National Defense University of Taiwan. In an article on the "Military Ethics of Xunzi," published in *Comparative Strategy* (Yu 2016), Yu argues that Xunzi's view of the ethics of war is significantly different from that of Mencius.

Regarding the views of Mencius and Confucius, which Yu refers to as "traditional Confucianism," Yu's position is, in broad strokes, similar to what I have described. He writes, "Traditional Confucianism opposes using war to solve problems because violence cannot resolve violence, but rather creates hatred and results in continual confrontations" (Yu 2016, p. 265).[2] Instead, traditional Confucianism supports "ethical governance" which involves "enlightening people through the king's benevolent personality and moral deeds, and achieving world peace through human[e] and ethical governance" (Yu 2016, p. 266). He further explains, "the governance strategy proposed by

Mencius denotes the process of inward cultivation to outward governance, which originates from a person, influences others gradually, and ultimately causes world peace" (Yu 2016, p. 263).

So far so good. But Yu interprets Xunzi as having quite a different view from that of Mencius and Confucius. Suggesting that Xunzi maintained that "One must prevent violence and eradicate evil through war" (Yu 2016, p. 260), Yu writes, "Xunzi promoted the use of temporary military actions as a means for restoring social order, changing the traditional Confucian antiwar perspective of 'those who are benevolent have no enemy' advocated by Confucius and Mencius" (Yu 2016, p. 260).[3] Yu explains:

> Xunzi maintained that when states are hostile toward each other, not only is implementing the ethical conscience strategy proposed by traditional Confucianism too slow for critical situations, but changing social circumstances is also impossible because humans are inherently evil. Only by maintaining social order through war can a state provide a stable living environment and cultivate the people through the environment. (Yu 2016, p. 266)

Yu's interpretation seems to be based more on inferences from an exaggerated version of Xunzi's view of human dispositions than on a close reading of Xunzi's comments about the use of the military. He suggests that because "humans are inherently evil" the optimism of Mencius and Confucius regarding the moral development of people is misplaced. For moral development depends on the proper environment, which, Yu reasons, can sometimes only be brought about by war. He writes, "the environmental determinism proposed by Xunzi provides legitimacy for war" (Yu 2016, p. 269). He explains, "By restoring social order through war and by providing people with stable lives, Xunzi aimed to transform human nature. Therefore, Xunzi adopted war as a means for maintaining social order, and believed that war was an inevitable behavior for fulfilling humaneness and justice" (Yu 2016, p. 271).

Recognizing that "the violent behaviors of war violate the Confucian values of humaneness and justice" (Yu 2016, p. 271), Yu argues that, for Xunzi, "humane and just wars aim to resolve the contradiction between violent actions and ethical governance" (Yu 2016, p. 271). However, while using violence to oppose violence may conceivably be justified in some circumstances, that fact does not resolve the means-ends conflict involved. One must simply judge that, in some circumstances, the ends *do* justify the means. But what are those circumstances, on Yu's interpretation of Xunzi? Yu writes, "Although Xunzi did not oppose war, he posited that it should be waged under careful consideration and only when absolutely necessary" (Yu 2016, p. 270). The phrase "absolutely necessary" is vacuous. Almost nothing is *absolutely* necessary. Actions can only be necessary *given certain goals*.

Perhaps the idea is that violence is sometimes necessitated by *ren* and *yi* (benevolence/humanity and a sense of appropriateness/justice). But what are the conditions under which this would be true? Yu suggests that for Xunzi, "foreign aggression or social unrest justifies violent behaviors (e.g., war) to maintain national security and social stability" (Yu 2016, p. 270). He also suggests that war is sometimes necessary to "maintain peace." He asserts that Xunzi "recognized war as a necessary evil to maintain peace while advising caution but supporting full employment of war when necessary" (Yu 2016, p. 270). However, he does not provide directly relevant evidence or elaboration.

One wonders what is meant by "full employment" of war. Regarding what is justified *in* war, Yu writes "Because the purpose of war is to satisfy social and public demands, everything that enables winning a war is legitimate" (Yu 2016, p. 267). And yet he also acknowledges that, for Xunzi, "military force should target the opposing army while protecting the common people, property, and crops of the enemy state" (Yu 2016, p. 269). These positions potentially conflict, and the desire for justice and humaneness does not resolve the conflict. Would Xunzi advocate the slaughter of some of the common people if that would provide an advantage resulting in the relief of a greater suffering. It seems Xunzi's answer is "no" (as Yu admits) even though Yu's analysis suggests that it should be "yes." Further, Yu also acknowledges that Xunzi imposes some serious limits on the legitimacy of the initiation of military force. Yu writes:

> If military actions cannot provide people with stable lives or if they endanger social order or people's lives, then they become illegitimate. Therefore, in principle, if the enemy is determined to defend a city, military leaders should not command soldiers to seize the city by force, to avoid unnecessary casualties; if the enemy forms a solid union with high morale or if the morale among [one's own] soldiers is low, a military leader should avoid dispatching troops on missions to preserve military strength. (Yu 2016, p. 269)

Further, Yu acknowledges that "Xunzi believed that a state should not force other states to surrender by military power even if the state possesses strong combat capabilities. Once a war has been initiated, other states must defend themselves, and their people are inevitably harmed" (Yu 2016, p. 265). And he also acknowledges that "Xunzi indicated that the success of a military action depends on unanimity among the people" (Yu 2016, p. 266).

Yu's account is deeply ambiguous. While he acknowledges grounds for interpreting Xunzi as having an extremely cautious attitude toward the use of the military, he nevertheless strongly implies that Xunzi has a rather permissive attitude toward war: War is justified by foreign aggression, unrest, national security, social stability, and presumably *any* circumstance in which concern for others can plausibly be asserted. In this simple way, Xunzi supposedly "reconciles the opposition between violence and the concepts of humaneness and justice" (Yu 2016, p. 261). But before we accept this, let's take a closer look at the *Xunzi*.

THE *XUNZI* ON WAR AND HUMANITARIAN INTERVENTION

Xunzi is asked why he consistently emphasizes a sense-of-humanity (*ren* 仁) and appropriateness (*yi* 義) when discussing military matters despite the fact that, according to the questioner, "Ordinarily, to make use of the military is to create conflict in order to take something by force" (*Xunzi* 15/71/22; 15.2).[4] In answering, Xunzi seems to give an argument for humanitarian intervention, which I will quote at length:

> Your understanding is not correct. Persons of genuine-humanity (*ren*) love people. Thus, they hate it when others harm them. Appropriateness (*yi*) involves according with practical coherence (*li* 理). But it is because they strive for practical coherence that they hate the chaos created by others. When they use the military, it is to stop violence and remove harm, not to create conflict in order to take by force. Thus, the use of the military by persons of genuine-humanity (*ren ren* 仁人) [is as follows]: where present, it is miraculous (*shen* 神); where passing through, it is transformative. Just as with the fall of the seasonal rains, all are delighted. This is why [the great sage] Yao toppled Huan Dou [and likewise with similar cases].[5] All conducted military activities throughout the civilized world with humanity (*ren*) and appropriateness (*yi*). Those nearby loved their goodness. Those far off longed for their appropriateness (*yi*). *Their armies did not bloody their swords*, for both far and near willingly submitted. (*Xunzi* 15/71/22; 15.2, emphasis added)

Although this passage does suggest that humanitarian interventions may sometimes be justified, it implies that it is only "persons of genuine-humanity (*ren ren* 仁人)" who may legitimately engage in such interventions. Xunzi makes this explicit earlier in the same chapter: "The *dao* [of deploying the military] of which I humbly refer involves the use of the military by persons of genuine-humanity (*ren ren* 仁人)" (*Xunzi* 15/68/11; 15.1b). Such persons are exceedingly rare. Confucius claims never to have met anyone truly fond of *ren* (*Analects* 4.6), and he does not even consider himself to be *ren*, or to qualify as a sage (*Analects* 7.34).[6] Mencius suggests that a sage can be expected only about once every five hundred years (*Mencius* 7B38). And that estimate is based on his dubious understanding of history. Modern Confucians should be even more pessimistic.

Further, the passage in question (*Xunzi* 15.2) suggests that force is justified only when there is virtually no resistance, since "all are delighted." In such cases, not much force would be needed. However, as I have stressed, reality does not present such cases. Xunzi has set an impossibly high standard. It would be, in Xunzi's own words, "miraculous" (*shen* 神) for this standard to be met. This is true even if we take the phrase "armies did not bloody their swords" as hyperbole, suggesting merely that there was not very much bloodshed. In the real world there is nearly always resistance to military interventions—resistance that would require significant killing to overcome militarily, with unavoidable "collateral damage."

The above-cited passage is not the only one in which Xunzi implicitly suggests very restrictive criteria for justified military action. In the passage immediately preceding it, Xunzi writes:

> With a true king (*wang* 王) there are punishments, but no battles. When cities are defended, he does not attack. Against a well-formed army, he does not strike. . . . He does not butcher the inhabitants of an [annexed] city, does not conceal his forces, does not engage in mass arrests, and does not command military service for more than one season. Thus, those who had suffered chaotic rule delight in his governance. And those who are insecure under their current leadership long for his arrival. (*Xunzi* 15/71/17; 15.1f)

In a note to this passage, Eric Hutton observes, "most commentators take Xunzi's point to be that a true king will not pursue a fight that would involve heavy casualties" (Hutton 2014, p. 368).

Accepting the key Confucian condition for appropriate military intervention—lack of significant resistance—but rejecting mythical accounts of history, contemporary Confucians should resist calls for military interventions and instead call for long-term Confucian-strategic efforts to live up to high moral standards, and thus heighten the moral standing of their country, and in turn strengthen its real and positive influence, even if it means forgoing an attempt at a violent quick fix.

Order, for Xunzi, is closely associated with peace and security. And Xunzi emphasizes that the key to order is the presence of persons exhibiting genuine-humanity (*ren ren* 仁人). "The virtue and generosity [of persons exhibiting genuine-humanity (*ren ren* 仁人)] are sufficient to pacify the common people, and the resonance of their virtue (*de* 德) is sufficient to transform them. If such people are obtained, there will be order; if not, there will be chaos" (*Xunzi* 10/43/20; 10.5). While this passage refers to domestic policy, its implications for international relations are not hard to surmise, as the dynamics that achieve peace in one's own state play out similarly internationally. Indeed, according to Xunzi, a true king (*wang* 王), on account of his "genuine-humanity (*ren* 仁), appropriateness (*yi* 義), and majesty (*wei* 威)," is esteemed not only by his own people, but universally. And so, Xunzi explains, "[A true king] is unopposed. And by virtue of the *dao* of protecting and serving the people, he is victorious without battles (*zhan* 戰), and gains [territory] without attacking, for the whole world submits without armor and weaponry being put to work" (*Xunzi* 9/37/15; 9.9; cf. 11.8).

Xunzi spells out the opposite dynamic in detail in *Xunzi* 9.7, below. There we see Xunzi reasoning about how the casualties of war, on all sides, lead to diminishing influence of the leaders who send their armies into battle. And this leads ultimately to ruin. The passage implicitly supports the notion that military engagements, for any purpose, should not be undertaken if there is significant resistance.

> Those who would use might: If city walls in other states are defended, and their people emerge to do battle, and I use my strength to defeat them, then there will surely be a great number of wounded among the common people of my adversaries. If so, they will surely detest me intensely. When other people detest me intensely, day-by-day they will grow more eager to fight with me. [In addition,] if city walls in other states are defended, and their people emerge to do battle, and I use my strength to defeat them, then there will surely be a great number of wounded among my own people. If there are a great number of wounded among my own people, then day-by-day they will grow less eager to fight for me. If others day-by-day grow more eager to fight with me, and my own people day-by-day grow less eager to fight for me, this is turning might into weakness. Territory is gained, but the people are lost. Exhaustive efforts are many, but achievements are few. Although what one defends increases, that with which one defends it decreases. This is how greatness is turned into ruin. (*Xunzi* 9/36/23; 9.7)

True strength, Xunzi goes on to explain, comes from virtue (*de* 德).[7] Xunzi says explicitly that this strategy will work in the absence of a sage-king, so we can be sure that this analysis is not limited to an ideal case.

Rather than criteria for the just use of violence, the Confucian focus is on developing a strategy for peace that has moral excellence, benevolent government, and a focus on ritual propriety as its root. First and foremost, the ruler should exhibit genuine-humanity (*ren*) (*Xunzi* 15.1b, quoted above). Such a person, rather than focusing on military matters, would achieve a well-governed state by emphasizing civil matters, ritual propriety,[8] and appropriateness.[9] Xunzi writes, "[A true king] could surely defeat [his opponent] in battle (*zhan* 戰), but would be ashamed to participate in fighting. He unassumingly achieves civil accomplishments for the world to see, and aggressive states peacefully transform themselves" (*Xunzi* 7/25/22; 7.1; cf. 25.28).[10] As a result of his civil accomplishments, he achieves overwhelming support of the people,[11] and that is the key to security.[12] Once this is fully accomplished, which is very rarely, punishing tyrants *without genuine war* becomes possible. Xunzi says, "The military forces of true kings are not put to the test. In Tang's and Wu's punishments (*zhu* 誅) of [the tyrants] Jie and Zhou, they [merely] clasped their hands, bowed, and then gestured a command with their fingers. And none of the mighty and aggressive states failed to hurriedly enable it. Punishing Jie and Zhou was like punishing solitary individuals" (*Xunzi* 15/70/14; 15.1d).[13] If one treats this as implying criteria for just military intervention, it seems that unless one is able to achieve victory without engaging in real battles, intervention is not justified.

XUNZI AND THE APPLICATION OF THE CONFUCIAN APPROACH IN THE CONTEMPORARY WORLD

Although it is beyond the scope of this chapter to address the empirical details of hard cases, I will offer one not-so-difficult example. While the "War on Terror" is not a paradigmatic case of humanitarian intervention, it was justified in humanitarian terms, namely, to defeat terrorists and thus save their innocent potential victims. But using violence against violence, according to a Confucian way of thinking, would only encourage more violence, as the family members of victims of wayward bombs and ill-fated drone strikes vow revenge. Indeed, the war on terror has not made Americans, or anyone else, safer. Imagine that, instead of waging war, half of the enormous monetary costs of those efforts were channeled into sincere and sensible efforts to improve the conditions of suffering people around the world by providing basic necessities, inexpensive mosquito nets, low-cost manual water pumps, low-interest micro-loans, and so on. Would that not have been more effective at reducing terrorism, and promoting a spirit of good will and friendly feeling toward the United States? That would have been the genuinely humane, and more effective, thing to do; that is the Confucian solution. Xunzi writes:

> Genuine-humanity (*ren* 仁), appropriateness (*yi* 義), virtue (*de* 德), and conducting oneself (*xing* 行) accordingly is characteristic of the style that reliably achieves security (*an* 安), even though the possibility of danger cannot be fully avoided. . . . Thus, exemplary persons lead (*dao* 道) according to what is reliable. Petty persons lead according to what is anomalous. (*Xunzi* 4/15/4; 4.8)

To repeat Tongdong Bai's remark, "the best way to disarm a threat and prevail in the world is to become a moral exemplar" (Bai 2012, p. 48).

Despite these observations, many people will likely continue to believe that there may be real conditions under which humanitarian military intervention would indeed be appropriate, especially when the carnage resulting from not going to war seems likely to dwarf that of going to war. But *how often* is that condition met? And, even if it is occasionally met, if appeal to humanitarian intervention is *more likely to be misused* than to be used appropriately, is it wise to accept this as a genuine moral option, without attaching unusually difficult-to-meet standards, as Xunzi has done? After all, Xunzi advises us to follow generally reliable paths. He writes, nearly echoing the passage cited above, "Exemplary people proceed (*dao* 道) according to what is regular, while petty people reckon what might be exploited" (*Xunzi* 17/81/1; 17.5).[14] The norms of ritual propriety embody "what is regular," that is, what is appropriate in general. These norms are, Xunzi writes, "the foundation of a strong state" (*Xunzi* 15/72/9; 15.4; cf. 10.13). This certainly applies to a state's security, for Xunzi writes: "Solid armor and sharp weapons are not sufficient to be victorious. High walls and deep moats are not sufficient to be secure" (*Xunzi* 15/72/10; 15.4). But the implications also seem more general. "If one follows the way (*dao* 道) [of ritual propriety], one will succeed (*xing* 行). If one does not, one will fail" (*Xunzi* 15/72/11; 15.4). (This is reminiscent of a point made in chapter 3, namely, that breaking norms for some greater good is not going to work.)

Xunzi twice appeals to the metaphor of norms of propriety being markers that enable people to ford rivers without falling into trouble (*Xunzi* 17.11, and 27.12). Here he makes clear that one reason to stick to regularities is the need for keeping the markers clear.

> Those who ford rivers indicate the deep spots. If the indications are not clear, people will fall in and get stuck. Those who bring order to the common people indicate the way. If the indications are not clear, there will be disorder. Norms of ritual propriety are the indicators. If ritual propriety is rejected it will be a dark age, and in such an age there is great upheaval. (*Xunzi* 17/82/22; 17.11)

Why would failing to be strict about observing norms of proper conduct lead to a dark age of great upheaval? That may seem rather extreme. But violations of norms have a way of escalating. And when violations become violent things can quickly get out of hand, even if one feels justified in one's violence. After all, it is not terribly difficult to imagine how a seemingly justified war might nevertheless engender animosities that spiral out of control (cf. *Mencius* 7B7). We cannot know how all the complexities will sort out.[15] So, in Xunzi's view, we should remain conservative and not assume we can reason out a better way than following peaceful principles.[16]

Further, even if significant suffering in particular locations might sometimes be reducible by means of military intervention, it is undeniable that significant suffering in many other locations may also be reduced by concerted non-military efforts (such as those mentioned above). The notion that any prosperous nation in the world is doing all it peacefully can to reduce serious and widespread suffering around the world cannot be seriously maintained. So long as a state could do more to help people peacefully but does not, how can it claim the moral authority to help people with violence?

In addition, when considering the potentially undesirable effects of resorting to military force to address problems, we should be mindful that questions regarding the use of the military go beyond the question of *under what conditions* it is justifiable to use force. For the very maintenance of a system that flexes its ability to use coercive force may have unwanted effects. In a chapter on strengthening the state, Xunzi explains the problems associated with reliance on military force (quoting at length):

> "*The method of force should cease, and the method of yi (appropriateness) should proceed.*" Why is this said? I say: it refers to the state of Qin. Qin's awesome military might rivals [that of the states formerly ruled by the sage-kings] Tang and Wu, and its lands are as vast as those of the sages Shun and Yu. Yet it cannot conquer its own anxiety. Consumed by apprehension, Qin is in constant fear that the whole civilized world will unite to crush it. This is what it means to say, "The method of force should cease." . . . So, what should Qin do? I say: *It should moderate its military might, and stress civilian issues instead.* It should then utilize exemplary persons who are upstanding, honest, reliable, and complete to bring order to the civilized world. Availing themselves of this contribution to the state's governance, they should properly align right and wrong, regulate the crooked and the straight, and put the capital (Xianyang) in order. (*Xunzi* 16/77/16; 16.5, emphasis added)

This passage reinforces the main point that one's focus should be on being morally upright oneself and on ensuring that one's own state is well-governed. Only once one has been successful at this—a project that is never complete—is one qualified to use force to correct others. In the meantime, as one makes steps in the right direction, one becomes an increasingly positive influence in the world. This is not callously leaving others to suffer. It is doing the best that can be done.

NOTES

1. In addition to the article by Yi-Ming Yu, discussed in the main text, Arron Stalnaker also explores some implications of Xunzi's views on warfare, in an article titled, "Xunzi's Moral Analysis of War and Some of its Contemporary Implications" (2015). However, Stalnaker does not provide significant textual evidence to support a permissive interpretation. Instead, he defers to the arguments given by Twiss and

Chan, who discuss the views of Mencius and Xunzi together. For example, Stalnaker remarks, "As Sumner B. Twiss and Jonathan K.L. Chan have shown, Xunzi and Mencius. . . share a very similar account of justice in war (see [Twiss and Chan 2015a]). They both condemn aggressive wars for territorial expansion, but permit and even encourage wars of self-defense against aggression, and also what they call 'punitive expeditions' by rightfully authorized state actors to redress unjust attacks against other states" (Stalnaker 2015, p. 135). Note that Stalnaker's comment suggests that it is only "rightfully authorized state actors" who may legitimately engage in such expeditions. I discuss Twiss and Chan's interpretation in chapter 6, below, including the question of who counts as a legitimate authority in this context.

2. Yu elaborates, "Traditional Confucianism indicated that using coercive means to resolve disputes or launching wars to settle international conflicts may easily cause eternal societal confrontation resulting in a cycle of murder based on revenge" (Yu 2016, p. 265).

3. Similarly, Yu also writes, "[U]nlike the negative attitude of traditional Confucianism toward war, Xunzi maintained that war is a necessary means to restore social order. According to this concept, war is no longer in opposition to morality but an unavoidable step for solving practical problems" (Yu 2016, p. 267).

4. There are several points at which Eric Hutton's translations of the *Xunzi* differ from Knoblock's in ways that are relevant to my thesis. It is worth noting a couple of significant discrepancies. Hutton translates a line in *Xunzi* 18.3 as follows: "Carrying out punitive campaigns against those who are violent and executing those who are brutal is the blossoming of order" (Hutton 2014, p. 188). Putting it this way invites an interpretation like that of Ni Lexiong, glorifying war. On the other hand, Knoblock's translation is very different: "To correct violent behavior with punishment and rebuke the cruel is the fulfillment of good government" (18.3). Elsewhere, Hutton uses the word "attacking" (Hutton 2014, p. 162) where Knoblock uses the phrase "set out on a campaign" (15.6b). Knoblock's language reflects alertness to the critical distinction between moving one's forces and actually fighting. There is a very similar discrepancy between Hutton p. 116 and Knoblock 11.13b. For other issues involving Hutton's translation, see notes 10 and 11 below.

5. The other examples are: "Shun toppled the rulers of the Miao, Yu toppled the Gonggong, Tang toppled the ruler of the Xia dynasty [the infamous tyrant Jie], King Wen toppled Chong, and King Wu toppled Zhou [the last ruler of the Shang dynasty]."

6. Confucius does say, however, that his protégé, Yan Hui, could go three months without deviating from *ren* (*Analects* 6.7). Unfortunately, he died young, after living contentedly in abject poverty. He was the farthest thing from a ruler. So, as a near-exception, he was an exception that proves the rule. Namely, rulers who would qualify as persons of genuine-humanity are rare indeed.

7. Xunzi also quotes an ancient saying, "The exemplary person uses virtue/potency-of-character (*de* 德); the petty person uses physical-power (*li* 力)" (*Xunzi* 10/44/5; 10.6). A similar distinction occurs in *Xunzi* 15.6a. Xunzi again argues, on the one hand, "One who uses physical-force (*li* 力) will be weakened" (*Xunzi* 15/74/16; 15.6a). And, on the other hand, if one uses moral-power, one will become a true king.

Xunzi writes, "If the people esteem one's fame and reputation, and admire (*mei* 美) the application of one's potency-of-character, and desire to be one's subjects, then they would welcome one's entry with open gates and cleared roads. If one allows the common people to continue [their practices] and carry on living in the same place, and one leaves all the people in peace, they will abide by the laws and carry out one's commands, without anyone failing to be submissive. In this way, one's authority increases more and more as one gains territory, and uniting the people [under one's rule] makes one's military strong. This is what comes of uniting people using virtue" (*Xunzi* 15/74/9; 15.6a).

8. Xunzi does suggest that ritual and music are incorporated into military movements and displays in order to "provide form to awe-inspiring majesty" (*Xunzi* 19/95/14; 19.7b; cf. 20.1). But genuine warfare is another matter. Xunzi quotes a saying, "Let your might be awe-inspiring, but do not make use of it" (*Xunzi* 15/73/9; 15.4).

9. Xunzi stresses that a well-ordered state will be a strong one, and that the keys to order are moral leadership, exalting ritual propriety (*li*), and valuing appropriateness (*yi*). Tactical matters, on the other hand, are of secondary importance (*Xunzi* 15.1c; cf. 15.3, 9.18, 10.13 and 16.4).

10. The passage continues: "Only those who have been catastrophically perverse are executed (*zhu* 誅). Thus, executions by a sage king are extremely rare." The character *zhu* 誅 means punish/kill, and in this context may imply "punitive expedition." And so, Eric Hutton renders the first sentence as follows, "Only when there are disastrous or perverse states does he launch a punitive campaign" (Hutton 2014, p. 48). In any case, it should be understood that this would not involve significant fighting, in which a true king would be ashamed to participate, according to this passage. Also, the extreme moral differential between the sage king, whose civil accomplishments are clear to all, and a "catastrophically perverse" tyrant should make genuine war unnecessary. Of course, these fanciful conditions are "extremely rare."

11. In Hutton's translation, Xunzi says, "[I]n the way of the ancients, the fundamental task for all use of military forces and offensive warfare lies with unifying the people" (Hutton 2014, p. 145). Out of context, Hutton's translation may be misunderstood to suggest that offensive warfare was acceptable *for the purpose of* unifying the people (though context suggest this is not Hutton's intention). Knoblock's translation makes it clearer that this passage is actually stipulating that the *prior unification* of the people is a requirement: "[In] the way of the Ancients, it was a general principle that the fundamental requirement to be met before using the army in attacks and campaigns was the unification of the people" (15.1a). At the end of this section Xunzi emphasizes the critical importance of popular support, "The essential point regarding the military is the support of the people, and that is all" (*Xunzi* 15/68/8; 15.1a).

12. How does one achieve a secure state? Xunzi's answer is surprising: "How a person who has humanity (*ren*) uses the state: he cultivates aspirations and intentions, assures proper personal conduct, exalts what is illustrious and lofty, engenders the utmost in faithfulness and honest commitment, and elevates refined cultural patterns to the utmost" (*Xunzi* 10/48/9; 10.14). How does this achieve security? Xunzi's explanation is as follows. Those who launch wars do so for the sake of good reputation,

or for profit, or else out of anger. But the person of genuine-humanity provides no excuses for anger. And, by virtue of his popular support, any assault against him will result neither in profit nor improved reputation. For the people will seize the enemy's generals as easily as "stirring cooked wheat."

13. Cf. "With a reputation sufficient to burn them, and awe-inspiring strength sufficient to trounce them, they [merely] clasped their hands, bowed, and then gestured a command with their fingers. And none of the mighty and aggressive states failed to hurriedly serve them. [Dealing with such states] would be like [the strongman] Wu Huo taking hold of a dwarf" (*Xunzi* 10/49/6; 10.15).

14. Cf. "Exemplary people (*junzi* 君子) know that what is neither whole (*quan* 全) nor pure (*cui* 粹) is not sufficient to be deemed beautiful or admirable (*mei* 美). Thus, by recitation and following regular patterns (*shu* 數) they habituate themselves to it [i.e. the beautiful/admirable]" (*Xunzi* 1/4/16; 1.14).

15. As Tim Hayward, quoting Henry Shue, points out, "The crucial question regarding the possible justification for any kind of intervention—preventive or humanitarian—is whether intelligence 'can be sufficiently reliable that it is not irresponsible to take human lives on the basis of it'" (Haward 2019, p. 541; quoting Shue 2009, p. 316).

16. Unlike the hypothetical cases often used in support of just war theory, Richard Jackson points out, "in the real world, wars are messy, complicated affairs that confound attempts at clarity and certainty" (Jackson 2020, p. 47). I would add that this also contrasts with the idealized semi-mythic "history" that Mencius uses to justify the deeds of certain ancient sage kings. In actuality, Jackson notes, "so many real world humanitarian interventions launched on the justification of JWT [Just War Theory], such as the interventions in Somalia, DRC, Haiti, Syria, and elsewhere, faced resistance and created unpredicted results, such as the violent instability following the Libyan intervention in 2011" (Jackson 2020, p. 49).

Chapter Six

Mencius and Xunzi on Tyranny and Humanitarian Intervention: A Response to Twiss and Chan

This chapter continues the discussion of war and humanitarian intervention, focusing on works co-authored by Sumner Twiss and Jonathan Chan, who treat both Mencius and Xunzi together. In their interpretation, Mencius and Xunzi imply that tyranny or "serious misrule" justifies humanitarian military intervention. Twiss and Chan highlight passages that might be thought, at first blush, to support a relatively permissive position regarding punitive expeditions. But on closer inspection, I argue, they do not.

TYRANNY AS JUSTIFICATION FOR MILITARY INTERVENTION

Twiss and Chan interpret Mencius and Xunzi to be "reasonably similar in their views" regarding the legitimate use of military force and so treat them largely as having "a unified position" on this matter (Twiss and Chan 2015a, p. 93). They provide evidence that putting a stop to severe tyranny is, for Mencius and Xunzi, a legitimate motive for engaging in a punitive expedition. Although they recognize other familiar just war criteria, they frame them so that they are relatively easy to achieve, and place significant emphasis on tyranny, or "serious misrule," treating it as one of "the threshold conditions or criteria that trigger the true king's obligation to undertake a punitive expedition against a ruler within the empire" (Twiss and Chan 2015b, p. 124). They say that Mencius and Xunzi are both "utterly clear" about this. I argue that, on the contrary, it is misleading to suggest that Mencius and Xunzi maintained that foreign tyranny requires military intervention.

It should be noted that Twiss and Chan aptly framed their question as addressing the obligation of a *true king*. I have argued (pp. 71–72, above) that, for a ruler who adequately satisfied the attendant moral requirements, certain conditions would be apparent and those can be viewed as highly restrictive criteria. However, Twiss and Chan don't treat these conditions as seriously restrictive. And despite the way they have framed it in the above quotation, they don't even actually maintain that the tyranny-trigger applies only to true kings, as I discuss below. In any case, the main point *here* is that, while removing tyranny does seem to be a legitimate Confucian *motive* for engaging in a punitive expedition, and maybe a necessary condition, that does not mean it is sufficient on its own to legitimate such an expedition. There are other conditions that must be met. As Sungmoon Kim puts the same point:

> [T]he presence of the oppressed people may offer a necessary condition for punitive expedition, but by no means can it be a sufficient condition. It may motivate the benevolent ruler to be concerned with the well-being of the people who are suffering from a tyranny, but it can never (fully) justify his decision to launch a war of punishment. (Kim 2017, p. 202)

Twiss and Chan do suggest that familiar Western just war criteria also apply in early Confucian thinking: just cause, right intention, last resort, reasonable chance of success, and proportionality (2015b, p. 128).[1] However, according to Twiss and Chan's interpretation, tyranny, and the intention to respond to it, satisfy both of the first two criteria (just cause and right intention). The additional requirement of a "reasonable chance of success" is not significantly restrictive. And Twiss and Chan seem to suggest that the requirements of proportionality and last resort are frequently met. However, if there is going to be significant resistance and thus genuine war is going to transpire, then proportionality is difficult to judge in advance. As for evidence that "last resort" was operative in Mencius's thinking, Twiss and Chan cite an example in which King Tang "initially used means other than military force before resorting to it" (2015a, p. 102), which they acknowledge is weak evidence. Indeed, if that is the standard, then any effort, however feeble, toward resolving an issue diplomatically before engaging the military may pass as satisfying the "last resort" requirement.

What Twiss and Chan emphasize as providing justification for military intervention is what they refer to as "severe tyranny" and "serious misrule." In this regard, they repeatedly state that the Confucian threshold for justification for intervention is lower than Western models. For example, they write:

> [T]he Confucian understanding appears to set a threshold standard for military intervention somewhat different from, and lower than, both of the Western models [i.e. the traditional legalistic and the revisionary moral models]—that is, "severe tyranny" that may not involve massive human rights violations on the

scale of, say, genocide or ethnic cleansing. (Twiss and Chan 2015b, p. 127; cf. pp. 124, 129,130)

Twiss and Chan are not alone in focusing on misrule as a justification for intervention. Cao Qin also maintains that, "According to Mencius's standard, all states with a sufficiently bad government are legitimate targets of intervention" (Qin 2020, p. 165). And, at least with regard to Mencius, Qin also concurs with Twiss and Chan regarding the relatively low "triggering conditions" for humanitarian intervention. He suggests that "we can safely assume that the bar would be lower" for Mencius when compared to the current international consensus (Qin 2020, p. 165). He explains, "If we adhere to Mencius's idea of just war instead [of assuming that the principle of sovereignty should take priority], we will probably tend to be more tolerant to humanitarian intervention. For Mencius, the fundamental moral criterion for judging the justice of a war was the consequences it would bring" (Qin 2020, p. 165). As I argued at the end of chapter 3, on the contrary, Mencius was not a simplistic consequentialist. If one insists on using the consequentialist lens to view Mencius, he must be seen as evaluating long time frames and taking into consideration feedback systems. Such a view leads him, and Xunzi as well, to support the reliable way: rulers should generally follow the path of sound, benevolent governance, rather than resorting to violence when it seems, superficially, like there might be a sufficiently good outcome to counterbalance the sure evil involved in violent conflict. Though there may be hypothetical or idealized exceptions, real cases are at best rare. And even in those cases, justifiable confidence that it is truly such a case is unlikely to be high.

Though less horrific than genocide, Twiss and Chan's characterization of "serious misrule" nonetheless involves quite serious abuses and atrocities:

> [It includes] deprivation of people's livelihoods, separation of families, murder, misuse of the criminal justice system (e.g., impunity for perpetrators), barbaric treatment of individuals (e.g., torture), predation or stealing people's goods, terrorization of the population at large, and what is effectively internal chaos or anarchy. (Twiss and Chan 2015b, p. 124)[2]

The emotional appeal of their argument is that it would seem that Confucians, who are concerned about the wellbeing of people, would want to put a stop to such misrule. On the whole, their analysis gives the impression that Mencius and Xunzi had a relatively permissive stance regarding military interventions and suggest that they both regarded the existence of tyranny as often *requiring* a military response. They write, "[A] Confucian contribution comes in its emphasis on responsible sovereignty as entailing a right and duty to rectify a severely tyrannical situation" (Twiss and Chan 2015b, p. 131).[3]

Of course, nobody wants atrocities to be allowed to continue. Nevertheless, relevant additional questions include: On Mencius and Xunzi's account, what circumstances are required for one to have confidence that military intervention isn't going to do more harm than good? And, do Confucians have a strategy that is an alternative to military intervention? To the first question I have suggested that such confidence can only be justified in circumstances that are exceedingly rare in the real world. As for the alternative strategy, it is to "play the long game." That is, recognizing that effective quick fixes to tragic circumstances are sometimes just not available, one ought to "return to the root" by increasing one's own influence through moral self-cultivation and continually implementing measures that bring relief to those who can be nonviolently helped (see *Mencius* 1A7). Below, I consider the evidence cited by Twiss and Chan to see whether they've offered good reasons to revise these conclusions.

AN EXAMINATION OF THE EVIDENCE

Twiss and Chan make their arguments in two chapters of a recent book, *Chinese Just War Ethics*. Their first chapter covers Confucian criteria for the legitimate initiation of military force as well as the limits on how such force may be used. The second chapter more explicitly compares the Confucian notion of "punitive expedition" with the contemporary notion of "humanitarian intervention." Much of the textual evidence relevant to the justification of humanitarian intervention, however, falling under the rubric of "legitimate use of military force," appears in their first chapter. In the main text of that chapter, Twiss and Chan prominently quote from four passages (*Mencius* 1B3, 1A5, 5A7, and *Xunzi* 15.1f)[4] to illustrate that military actions to punish and rectify aggression and tyranny are regarded by Mencius and Xunzi as justified (see Twiss and Chan 2015a, pp. 99-100). Below I consider each passage in turn, beginning with the relevant quotation as it is given by Twiss and Chan. (They use D.C. Lau's translations for the *Mencius* and John Knoblock's for the *Xunzi*.)

Mencius 1B3

King Xuan of Qi asked Mencius advice regarding interactions with neighboring states, mentioning his fondness for courage. In response, Mencius encouraged the king not to be fond of *petty* courage, and cited King Wen and King Wu as examples to follow:

The Book of Odes says, "The King blazed in rage and marshalled his troops to stop the enemy advancing on Ju" . . . This was the valour of King Wen. In one outburst of rage King Wen brought peace to the people of the Empire. . . . If there was one bully in the Empire, King Wu felt this to be a personal affront. This was the valour of King Wu. Thus he, too, brought peace to the people of the Empire in one outburst of rage" (*Mencius* 1B3)

The meaning of this passage is not transparent from the passage itself. The passage actually ends with the following (still using Lau's translation): "Now if you [King Xuan of Qi],[5] too, will bring peace to the people of the Empire in one outburst of rage, then the people's only fear will be that you are not fond of valour." It is not plausible, however, that Mencius is encouraging King Xuan to burst out in rage and bring peace to the Empire. After all, in *Mencius* 1A7, Mencius warns him that (as Bryan Van Norden explains the situation), "given the military resources of the state of Qi, the consequences of trying to conquer all of the Middle Kingdom would be disastrous" (Van Norden 1997, p. 251; see pp. 69–70 above). Van Norden suggests the story in question (i.e., *Mencius* 1B3) can be understood in reference to the discussion of courage in *Mencius* 2A2. There Mencius suggests that the highest courage is a kind of equanimity that comes from nurturing one's qi with honorable conduct (*yi* 義) and uprightness. With this in mind, *Mencius* 1B3 may be interpreted as suggesting that King Xuan should focus his efforts on becoming truly honorable and upright, and thereby eventually become able to successfully accomplish, in a single bout, something akin to the successes of King Wen and King Wu[6]—both of whom would qualify as genuinely *ren* true kings. This will doubtless require a great deal of self-cultivation on the part of King Xuan. And in the meantime, the King is being urged to put aside his affinity for "petty valor." That's the point. It would be a mistake to view this passage as greenlighting punitive expeditions or humanitarian interventions by someone of less-than-exceptional moral stature.

Mencius 1A5

Having suffered a number of humiliating military defeats that has left his state smaller and weaker than it once was, King Hui of Liang asked Mencius for advice. Mencius answers:

These other princes take the people away from their work . . . making it impossible for them to till the land and so minister to the needs of their parents. . . . These princes push their people into pits and into water. If you [Your Majesty] should go and punish such princes, who is there to oppose you? (*Mencius* 1A5)

Taken out of context, this quotation is a bit misleading. It sounds as if Mencius is encouraging King Hui to engage in a punitive expedition. But that is not what is happening. First of all, the critical part is ambiguous. It could be read, "If a *true king* were to conduct punitive expeditions against them, who would treat him as an enemy?" But even if we stick with D.C. Lau's translation it is perfectly clear in context that what Mencius is doing here is encouraging King Hui to *govern more humanely himself.*

After all, Mencius criticizes the neighboring kings only after having implied that King Hui is himself guilty of *the very same offenses as they are.*[7] This suggests that if a true king were to overthrow *him* no one would object any more than they would object to seeing neighboring kings overthrown. Mencius is even implying that those military and territorial losses Hui is smarting from might have been deserved, meaning he has no right to expect them to be rectified. Mencius says, "A territory of a hundred *li* square is sufficient to enable its ruler to become a true King" (*Mencius* 1A5, Lau's translation). That is, if Hui wants to make things right, he should concentrate on ruling better rather than fretting over how much territory he rules over, and how he can win back what he had lost. What is more, as a practical matter, he will not have the cooperation from his own people necessary to mount a military campaign, or even an effective defense against further incursions by his neighbors, unless he first wins the loyalty of his own subjects by reforming his own economic policies to improve their quality of life.

Mencius says, "If Your Majesty practices benevolent government towards the people . . . then they can be made to inflict defeats on the strong armour and sharp weapons of Qin and Chu, armed with nothing but staves" (*Mencius* 1A5, Lau's translation). Van Norden's translation of the crucial line makes the import explicit: "If Your Majesty had applied benevolent government to Your own people, and then came to attack those other rulers, who would oppose You?" (Van Norden 2008, p. 6). Mencius's objective is to get the king to improve himself and institute more benevolent policies. And he concludes: "Thus it is said, 'Those who are *ren* have no enemies.'" According to Mencius, if punitive action is called for, those who are morally qualified to administer it could do so essentially unopposed. (See p. 74, above, for more analysis of this passage.)

It would therefore be a mistake to view this passage as implying a permissive view of humanitarian intervention.

Mencius 5A7

While recounting the career of Yi Yin, a legendary advisor to King Tang, Mencius mentions that

[In response to tyranny] he [Yi Yin] . . . went to Tang and persuaded him to embark upon a punitive expedition against the Xia to succour the people. (*Mencius* 5A7)

This passage might be taken to mean that rescuing people from tyranny is sufficient cause to wage war against the tyrant. Once again, though, the ruler who is regarded as rightfully engaging in a punitive expedition counts as a true king. When King Tang began his punitive expeditions, Mencius tells us in 1B11, "all the world trusted him," and thus his armies were welcomed. And Yi Yin himself was a moral exemplar and contributed to Tang's moral development. Mencius explains, "If it was dishonorable, if it was contrary to the way, though offered the whole world, he [Yi Yin] would not even consider it" (*Mencius* 5A7).

We have seen that Mencius sometimes appeals to idealized semi-mythic tales to contrast with the situations he faces. This case is different. It is not really about the use of the military. It is about Yi Yin not trying to achieve power by using his cooking skills to win favor with Tang. The point being that he could not have succeeded in facilitating the moral development of Tang if he had compromised himself. But for our purposes, the important thing is that Tang qualifies as a true king, who could, as Xunzi explicitly describes, eliminate the scourges of the world and have the whole world turn to him by cultivating the way and putting honorable conduct (*yi* 義) into practice (see *Xunzi* 18.2, quoted on p. 147 n23 below).

Xunzi 15.1f

Xunzi explains to King Xiaocheng and the Lord of Linwu,

> In the rule of the True King there are punitive expeditions but no [aggressive] warfare. . . . When the ruler and his subjects are pleased with each other, congratulations are offered . . . Thus those who live in anarchy rejoice in his government and those discontent with their own ruler [living under tyranny] desire that he should come (*Xunzi* 15.1f, Knoblock's translation).

Here the translation/interpretation of the first line makes a significant difference. I translate it as "With a true king (*wang* 王) there are punishments, but no battles." This makes sense of the following sentences, which Twiss and Chan skip over: "When cities are defended, he does not attack. Against a well-formed army, he does not strike." (See the discussion of this passage on p. 83 above.)

Once again, this passage only applies to the rule of a true king. And it implies that a military intervention would only be justified in cases in which there would not be significant military resistance. So, none of the four passages that were highlighted by Twiss and Chan truly support the notion that Xunzi and Mencius would be supportive of humanitarian interventions in today's world if they involved actual war rather than being more like a police action. It should be remembered too that they did not support any such military force in their own time and Mencius claimed that there were no just wars in the preceding three-hundred-year Spring and Autumn period. Their positive, generally contrastive, examples were all idealized semi-mythic tales told for another purpose, which usually involved encouraging a ruler to concentrate on governing his own people better.

Additional Passages

In addition to the above quotations, Twiss and Chan make reference to a number of other passages. For example, referring to *Xunzi* 15.2, Twiss and Chan write, "Xunzi is saying that there can be a legitimate use of military force insofar as such use serves benevolence and righteousness" (Twiss and Chan 2015a, p. 99). This suggests that right intention is a sufficient justification to wage war. But this interpretation ignores the fact that exemplary examples of the use of the military involve armies that "did not bloody their swords" (*Xunzi* 15.2, see pp. 82–83, above). That is, right intention, while it may be a necessary condition, does not seem to be a sufficient one. The exemplary cases suggest that the condition of non-resistance must also be met.

Also, while suggesting that "severe tyranny is the threshold standard for a true king to undertake a punitive expedition," Twiss and Chan quote *Mencius* 7B4: "To wage a punitive war is to rectify" (as cited on Twiss and Chan 2015b, p. 125). But they fail to mention that this sentence is followed by the concluding lines of the passage, which on D.C. Lau's translation reads: "There is no one who does not wish himself rectified. What need is there for war?" (See p. 76 n9 for my translation of the whole passage.) Far from setting a relatively low threshold for just warfare, the point of the passage is that by appealing to the higher aspirations of people, war can be avoided. (At least here Twiss and Chan make this applicable only to a "true king." But they gloss over the restrictive implications of that detail, and, as I discuss below, elsewhere argue for a looser restriction.)

Further, in an endnote, Twiss and Chan quote *Mencius* 6B7 (2015b, p. 132 n8),[8] seemingly to support the idea that serious misrule is sufficient to justify a "punitive military expedition in order to rectify the situation" (Twiss and Chan 2015b, p. 124). There are a couple of problems with this. First, in 6B7, it is not the horrors that characterize "serious misrule" that serve as a trigger, but rather "failure to attend court" three times.[9] Second, it appears that it is not a trigger to involvement in a serious military conflict, with all the death and destruction that predictably attends such actions. Twiss and Chan rely on D.C. Lau's translation, the relevant part of which reads: "If a feudal lord fails to attend court, he suffers a loss in rank for a first offence, ... and for a third offence the Six Armies will move into his state." But note that, on Van Norden's translation, the last bit reads, "the Six Bailiffs would remove him" (六師移之). Both translations are plausible, and both are consistent with the following point, though Van Norden's translation makes it more clearly: The passage does not support *war* motivated by humanitarian concerns. Indeed, the next sentence, even on Lau's translation, explains that "the Emperor punishes but does not attack." Presumably, extreme violence is not necessary because of the untenability of resistance that is assumed to apply in this case. After all, this example refers to the ideal relationship between the "Son of Heaven" (the Emperor) and a misbehaving feudal lord. The inapplicability of this feudal model to contemporary circumstances is another reason, one which Sungmoon Kim stresses (2017, pp. 188-192), that such examples are not really relatable to contemporary humanitarian interventions.[10]

Kim also argues that there is a distinction between *humane* intervention (intervention that results in increased human welfare) and *humanitarian* intervention (intervention *for the sake* of human welfare). Kim maintains that it is the former that Mencius condoned, and that is only *indirectly* related to tyranny: In Mencius's view, it is not actually the suffering of the people that justified punitive expeditions (Kim 2017, p. 202). It was rather the fact that the ruler had "gone astray," which the tyrannizing of the people simply demonstrates (Kim 2017, p. 197).

Kim makes the case that, according to Mencius, even though the people of Yan were suffering under misrule, King Xuan of Qi was not morally qualified to lead a punitive expedition to rectify the situation. He writes:

> [E]ven if the people's suffering in Yan was grave enough to require an external intervention, it is extremely difficult to identify who has the ritually sanctioned right to carry it out in the name of Heaven in the virtual absence of the institutional authority that used to represent the Mandate of Heaven vicariously. In Mencius's view, this practically baffles (and ought to baffle) any attempt by an ambitious ruler during the Warring States period to justify a punitive expedition. (Kim 2017, p. 191)[11]

Twiss and Chan acknowledge the essential point: "Mencius . . . denies that Qi's ruler has the legitimate authority to attack, since he was not a Heaven-appointed officer (that is, a true king with the Mandate of Heaven)" (Twiss and Chan 2015a, p. 107). However, as discussed directly below, Twiss and Chan do not hold firm to this standard.

WHO MAY CARRY OUT A PUNITIVE EXPEDITION?

True Kings

As I have repeatedly emphasized, an important criterion for the legitimate use of military force, for both Mencius and Xunzi, is the exceptional moral status of the leader who uses it. Sungmoon Kim seems to agree, at least with respect to Mencius. He makes the following point in this regard:

> Given the way in which Mencius's discourse of punitive expedition is constructed, centered as it is around the intervening ruler's moral virtue and Heavenly sanction, it is difficult to derive the motivation for and justification of humanitarian intervention directly from the massive scale of the people's suffering, which would make the military intervention in question truly a humanitarian, not merely humane, intervention. After all, it should be recalled, *Mencius did not endorse Qi's intervention in Yan, despite the fact that the people of Yan had indeed been undergoing unbearable suffering under the tyranny.* (Kim 2017, pp. 200-201, emphasis added)

Twiss and Chan ignore the emphasized point and seek to get around the strict moral requirement. At some points, however, they tacitly seem to acknowledge it. For example, they write, "when confronted by serious misrule, the true king has the rightful authority to undertake a punitive military expedition in order to rectify the situation" (Twiss and Chan 2015b, p. 124). So long as one is strict about the qualities of the true king, and what that entails in terms of the relatively non-violent use of force, I accept this. Regarding the relevant qualities of a true king, Twiss and Chan acknowledge: "The true king . . . is the humane and just ruler who has the Mandate of Heaven (as revealed in the people's support and loyalty) and who governs thoroughly and consistently according to the Confucian Way; he is the morally perfected ruler who governs and unites through the inherent power of his virtue" (Twiss and Chan 2015a, p. 97). In this context, they even quote the following passage:

> The Way of a True King is . . . like this. His humanity is the loftiest in the world; his justice the most admirable, and his majesty the most marvelous. His humanity being the loftiest is the cause of no one in the world being estranged from him. His justice being the most admirable is the cause of none failing to esteem him. His majesty being the most marvelous is the cause of no one in the world presuming to oppose him . . . [and] . . . coupled with a way that wins the allegiance of others is the cause of his triumphing without having to wage war, of his gaining his objectives without resort to force, and of the world submitting to him without his armies exerting themselves." (*Xunzi* 9.9, as quoted in Twiss and Chan 2015a, p. 98)[12]

This, as I have stressed, is a standard that nobody lives up to. If it is only a true king who has the moral authority to conduct punitive expeditions (since, on my reading, only under such a leader could the expeditions be accomplished with minimal violence), then, for all practical purposes, such expeditions are not permissible in the real world.

However, Twiss and Chan have a number of strategies for avoiding this conclusion. One is to recast true kingship as a more achievable status. In the following quotation, Twiss and Chan's characterization of true kingship makes it seem rather unremarkable; it involves benevolent governance, but in a mundane sense. They write:

> A true king has various role responsibilities: most especially, maintaining interstate order and peace within the empire and making sure that states are well governed. Good governance is focused entirely on the well-being of the people: that they and their families have decent livelihoods, are educated, have their physical security assured by the maintenance of ritual forms, a fair and competent criminal justice system, a social security net that supports them in times of distress or natural disaster, a system of fair taxation, a flourishing trading system, and the like. (Twiss and Chan 2015b, p. 124)

There is textual basis for each detail given; yet the overall impression is misleading. This can be seen by considering Mencius's comments quoted by Twiss and Chan in an endnote, which supports the various details:

> If you honour the good and wise and employ the able so that outstanding men are in high position. . . . In the market-place, if goods are exempted when the premises are taxed, and premises exempted when the ground is taxed. . . . If there is inspection but no duty at the border stations. . . . If tillers help in the public fields but pay no tax on the land . . . If you abolish the levy in lieu of corvee and the levy in lieu of the planting of the mulberry. . . . If you can truly execute these five measures, the people . . . will look up to you as to their father and mother. . . . In this way, there will be no match in the Empire for you. He who has no match in the Empire is a Heaven-appointed officer, and it has never happened that such a man failed to become a true King. (*Mencius* 2A5, quoted in Twiss and Chan 2015b, p. 132 n7, which follows D.C. Lau's translation).

The ellipses in the quotation above are Twiss and Chan's. To see what is misleading about their presentation, one need only fill in some of the gaps. When one does so, the last part becomes:

> If you can truly execute these five measures, the people *of your neighbouring states* will look up to you as to their father and mother; *and since man came into this world no one has succeeded in inciting children against their parents.* In this way, there will be no match in the Empire for you. He who has no match in

the Empire is a Heaven-appointed officer, and it has never happened that such a man failed to become a true King.

A true king, then, is someone who would be supported not only by his own people, but also by the people of neighboring states, *who would not fight against him*. Though *some* of the requirements Mencius mentions might seem easy enough to satisfy (abolishing import tariffs, for instance), Mencius's assertion that adequately satisfying them would make one truly exceptional—one would have no relevantly similar rival—suggests that it is unreasonable to think that contemporary leaders would somehow qualify.

Good Competent Political Leaders

Despite their less-than-awesome re-characterization of a true king, Twiss and Chan seem to sense that it would be a stretch to suggest that contemporary leaders could be credibly awarded this moral status. And they admit, "[E]ven at the time of their reflection and teaching, both Mencius and Xunzi were well aware that no true king then existed" (Twiss and Chan 2015b, p. 126). So, Twiss and Chan concoct a lesser category, a "good competent political leader" writing: "For both Mencius and Xunzi, second-best political leaders were those who followed the path of the ancient true kings, practiced humane governance, upheld justice, and ruled according to elaborate ritual forms" (Twiss and Chan 2015b, p. 126). Twiss and Chan admit that such political leaders did not measure up to true kings, which they here describe as being "endowed with complete virtues and talents and had the mandate of Heaven as well as support from all people" (Twiss and Chan 2015b, p. 126). They nevertheless suggest that, for Mencius, a true king's (ostensible) responsibility to undertake punitive expeditions against tyrannical leaders applies also to their newly invented category. They assert, "For Mencius, only a true king and a good competent political leader can be entrusted with the responsibility to interdict inter-state aggression and to respond to cases of severe tyranny by undertaking punitive expeditions" (Twiss and Chan 2015b, p. 126).

To motivate the category of a "good competent political leader" they cite *Mencius* 1A7, 1B1, 1B3, 1B5, and 2A1, though they do not explain exactly how these passages are thought to support the idea that Mencius employed this category in his thinking. On the other hand, *of course* rulers vary in quality, and so some rulers who do not measure up to a true king are relatively better than others and laudable in some ways or show signs of moral potential. But the critical question is this: Does Mencius suggest that these "second best" rulers have the moral authority to engage in punitive expeditions (or humanitarian military interventions)? Other than an enigmatic portion of

1B3 (discussed above), the passages cited in this context offer no evidence in favor of that dubious proposition.

Lord-protectors

Twiss and Chan suggest that while Mencius believes that "only" a good competent political leader and a true king may legitimately engage in punitive expeditions (and indeed have a duty to do so), Xunzi is even looser in his standards. Regarding a hegemon, or "lord-protector," who "governs his own state with a modicum of justice and humaneness, is non-aggressive toward other states, keeps his country strong, and keeps his word with friends, allies, and his own people," Twiss and Chan write, "such a hegemon can be entrusted, through the agreement of his allies, with the responsibility to interdict inter-state aggression and to respond to cases of severe tyranny by undertaking punitive expeditions" (Twiss and Chan 2015b, p. 126). This is despite the fact that, as Twiss and Chan acknowledge, lord-protectors began "using the pretext of protection and aid in order to intervene in, and gain advantage over, smaller states during their internal quarrels. As a result, ethical ideals were employed as a 'cover' by rulers to achieve their own aims, usually involving territorial aggression and expansion as well as internal tyranny" (Twiss and Chan 2015a, p. 94). In spite of this reality, Twiss and Chan point to Xunzi's relatively positive characterizations of lord-protectors as a basis for thinking that, in Xunzi's view, they could legitimately engage in punitive expeditions. Twiss and Chan argue:

> Xunzi also appears to regard a lord-protector as another sort of authority who has the right to undertake punitive expeditions: "He [the lord-protector] offers survival to those who face destruction . . . he guards the weak and forbids aggressive behavior . . . if there happens to be no True King ruling the world, he will invariably triumph. Such is one who knows the way of a lord-protector" (Twiss and Chan 2015a, p. 100, quoting *Xunzi* 9.8).

One problem with using this passage as evidence for their claim is that the passage appears purely descriptive, not normative. It only suggests such a ruler, being morally better than his adversaries, *will* triumph over them. It does not suggest they would be *justified* in undertaking punitive expeditions.

Similarly, referencing another passage, Twiss and Chan suggest that, for Xunzi, the lord-protector, though morally inferior to a true king "has nevertheless 'entered the precincts' of sufficiently rational government and order to be entrusted by other states with the responsibility to maintain it through punishment and rectification of aggression and tyranny" (Twiss and Chan 2015a, p. 101). They cite *Xunzi* 15.1d. However, once again, this passage merely suggests that lord-protectors *will be able* to defeat tyrants by virtue of their relative moral superiority. The considerations given relate to "signs indicative of strength and weakness." It does not suggest that morally imperfect lord-protectors may legitimately engage in punitive expeditions, much less that they have a responsibility to do so.[13] Rather, it suggests only that they will, *in fact*, triumph in battle over the less morally developed. Xunzi's claims here serve his purpose: to encourage everyone—even those whose aspirations are not pure—to develop themselves morally.

In the end, Twiss and Chan have not provided support that passes scrutiny for their position that early Confucians held that leaders exhibiting rather ordinary moral qualifications have either the right or duty to use military force against another state if its ruler is behaving badly. One more example, this one involving Confucius, is illustrative. In their discussion of the relevant history, citing *Analects* 14.16-17, Twiss and Chan write, "We propose—and this is our interpretation—that this status [i.e. lord-protector] gave Duke Huan the rightful authority to undertake punitive expeditions in order to save the empire and benefit its peoples" (Twiss and Chan 2015a, p. 113 n13). However, this pair of passages does not really even praise Duke Huan, much less imply that he had the moral authority to use violence justly. Instead, Confucius credits Duke Huan's successes to his advisor Guan Zhong's ability to unite people *without force*. While Guan Zhong was neither saint nor pacifist, it was his effective diplomacy that Confucius highlights in these passages.

THE NEED FOR OVERWHELMING SUPPORT

I have stressed that the legitimate use of force, for Mencius and Xunzi, is conditioned upon the overwhelming welcoming of the intervening force by the people of the intervened state, which is thought to be derivative of the exceptional moral status of the intervening authority.[14] In Twiss and Chan's brief critique of Bell, which in some ways coheres with my own,[15] they argue that the (rather weak) welcoming criterion proposed by Bell (see pp. 59–60 above) does not apply. For one thing, they claim, "In the case of the 'welcoming' criterion, Western discussions tend to tie this criterion to the right of peoples to their political self-determination, which, in the classical Confucian context, is simply inoperative" (Twiss and Chan 2015b, p. 128). But the people's wishes *do* seem to be operative in Mencius's thinking. As already mentioned, Mencius suggests that King Xuan should establish a ruler of a conquered territory "in consultation with the masses" (*Mencius* 1B11, see p. xiii above). And Mencius implies that the fact that an "attacking army is welcomed with baskets and kettles of food and drink" (*Mencius* 1B10, see p. 64 above) and that the people knock their heads on the ground in welcome are significant indications relevant to justification (*Mencius* 7B4, see pp. 61–62 above).

Further, Twiss and Chan suggest that the "welcoming" criterion, as well as the criterion involving the world's support,[16] may only be applicable "ex-post facto" and thus, they reason, "cannot be significantly operative in justifying at the outset the initiation of a punitive expedition" (Twiss and Chan 2015a, p. 111).[17] They argue that the people's "welcome" only functions post-intervention as a criterion for whether the already-intervened state can be legitimately annexed (Twiss and Chan 2015b, p. 129). Presumably they are thinking of the details of *Mencius* 1B10, in which this situation occurs (see p. 64, above). But as a general rule, this does not seem to be quite right. Signs of welcome, and the lack of resistance, may be evident at the time of attack (or of the last-minute decision not to attack). This reasoning seems evident in Xunzi's remark, mentioned above, "When cities are defended, [the true king] does not attack. Against a well-formed army, he does not strike" (*Xunzi* 15/71/17; 15.1f).

Twiss and Chan do note that Bell had suggested to them that "educated guesses" can be reasonably made about whether a punitive expedition would be welcomed and aptly pointed out that other criteria that Twiss and Chan regard as operative, such as estimating the likelihood of success, can only be educated guesses as well (Twiss and Chan 2015b, p. 132 n18). In response, Twiss and Chan write, "Bell's argument at best makes the 'welcoming' criterion a redundant one, because such educated guesses must be based on the benefits that will be brought by military action. In that case, such guesses would be taken care of by the principle or criterion of proportionality" (p. 132 n18). This does not seem right either. The welcoming criterion does not seem to be *entirely* subsumable under the criterion of proportionality. As I have indicated above, the wishes of the people *do* seem to be relevant to Mencius. Further, even if the "welcome" criterion were entirely based on its presumed practical effects, that would not mean it does not serve as a necessary criterion that cannot simply be outweighed by some other aspect of proportionality. Non-Confucians might be tempted to make all manner of guesses relevant to proportionality. But, it seems, for Mencius and Xunzi, the practical effects of universal welcome significantly change the moral calculus and may reasonably be thought to provide a clearer restraint on those tempted to resort to violence than does the more general notion of proportionality.

JUST WAR PACIFISM

Twiss and Chan briefly touch on the work of Julia Ching and P. J. Ivanhoe, who addressed the Confucian legitimation of war in connection to the question of the morality of weapons of mass destruction. Ivanhoe clarifies and agrees with the main point that Ching seemed to be driving at: in the

Confucian themes that Ching discusses "we can see a clear and respectable philosophical justification for rejecting any use of weapons of mass destruction" (Ivanhoe 2004, p. 270). Although it does not seem that Ivanhoe would go as far as I have in emphasizing the restrictiveness of the legitimate use of force from an early Confucian perspective, he does acknowledge this much: "The concept of punitive expedition employed by these early Confucians places fairly strong constraints on the nature of legitimate interstate conflicts. ... A ruler who could not liberate the people of another state without inflicting great harm on them would not engage in military action against them" (Ivanhoe 2004, p. 274).

In their brief but critical review of Ching's essay, Twiss and Chan write: "Ching claims that the early Confucians tended 'to be principled "just war pacifists," that is, they do not rule out war, but their criteria for a just war effectively limit (immediate or frequent) recourse to it'" (Twiss and Chan 2015a, p. 109, quoting Ching 2004, p. 253). In an endnote, Twiss and Chan comment, "It is not entirely clear how one is to understand the oxymoron 'just war pacifism,' nor is it obvious how classical Confucian thinkers who do hold a position on the legitimate use of military force are properly considered 'pacifists' in any serious sense at all" (Twiss and Chan 2015a, p. 115 n25). This comment is not entirely fair. "Just war pacifism" is only oxymoronic on a definition of pacifism that Ching clearly does not intend. And it is intelligible in a way that seems compatible with what Ching suggests: it implies a just war perspective that is so highly restrictive that all *actual* wars will count as unjust.

Although I'm not sure that I want to embrace the phrase myself, because of the ambiguity Twiss and Chan recognize in it, "just war pacifism" may be a reasonable way of characterizing the description of the early Confucian view of warfare that I have offered. I'm not sure there is a better summary term for the position occupied by Mencius and Xunzi alike. For *strict* just war reasoning of the kind I've described can be compatible with, and may even imply, a *kind* of pacifism. As Xinzhong Yao has commented, "Confucians argue for 'just war' only in a rigorously disciplined manner. Having permitted war only in a strictly ethical context, Confucians deny justice or rightness to all kinds of violence" (Yao 2004, p. 102). He explains, "[A true king] thus triumphs without having to engage in a fight, gains his objectives without resort to force, and submits the whole world to his rule without his armies exerting themselves. This is how a war can be won without killing and suffering" (Yao 2004, p. 101). This resonates strongly with the position I have argued for, though I would hesitate to call this "war."

In addition, it is important to recognize that there are many different versions of pacifism.[18] To claim the label "pacifist," one need not maintain that all violence of any kind is morally wrong in all metaphysically possible scenarios. A weaker form of pacifism may entail only that *offensive/invasive military violence* is morally wrong in all *plausible* scenarios *in our time*. It is just war impossibilism. I don't think Mencius and Xunzi went quite that far (in their time), for they seem to have held out hope for the emergence of a true king of outstanding moral quality. But in our time, I'm not sure even the most committed Confucians are entitled to *that* hope. As Ching comments, anticipating a theme I have emphasized: "[W]e should remember that the ancients attributed such punitive wars *to semi-legendary sage kings*. I do not think the Confucian tradition would support today's world powers in arrogating such prerogatives to themselves" (Ching 2004, p. 262, emphasis added). So, a contemporary Confucian who accepts the arguments given here may reasonably be thought to qualify as a "just war pacifist."

It should also be noted that there are a couple of ways in which early Confucian reasoning is closer to pacifist reasoning than to just war reasoning. Just war theory has a tendency toward "privileging the use of violence and then searching for reasons and occasions for its use"—at least according to Richard Jackson, director of the National Centre for Peace and Conflict Studies at the University of Otago, New Zealand (Jackson 2020, p. 46). Just war reasoning "starts with the question, under what conceivable circumstances is it legitimate for states to use violent force as a political means?" (Jackson 2020, p. 46). Certainly, that was not the Confucian approach. So, even if early Confucians did think war was sometimes justified, it can be misleading to treat them as just war theorists. In contrast, Jackson writes, "[A] pacifist approach starts by asking, what is the most effective and ethical thing to do in response to a threat in a specific cultural-historical context?" (Jackson 2020, p. 46) While Jackson is not writing about Confucianism, this sounds *precisely* like the Confucian approach. Confucians would say that dealing with threats requires sticking to the same program of effective and ethical governance that applies generally. They tend to think that the answers adhere to norms that facilitate productive social relations, admirable conduct on the part of leaders, and the institution of policies that improve people's lives. If Jackson's characterizations of just war theory and pacifism are accepted, it seems that early Confucians had more in common with pacifists than with just war theorists, *at least with regard to their fundamental approach to conflict solving.*

Further, pacifists commonly reject the suggestion that they advocate *passivity*. Instead, they advocate "engagement in the present to construct more just and peaceful societies so that violence does not emerge in the first instance" (Jackson 2020, p. 55). Mencius and Xunzi can both be seen as pacifists in this active sense, steering rulers away from violence toward engagement in more benevolent governance, both discouraging war in the present and creating conditions that reduce the likelihood of war in the future.[19] It can be considered a pacifistic strategy, even though the use of the military is condoned in idealized circumstances.

REFLECTIONS ON POINTS OF AGREEMENT WITH PING-CHEUNG LO

Together with Twiss, Ping-cheung Lo co-edited *Chinese Just War Ethics*, which contains the essays by Twiss and Chan under review here. Lo also wrote the first chapter, which provides an overview of the book. Therein, Lo seems to endorse Twiss and Chan's interpretation regarding the early Confucian perspective on the just use of the military, and on humanitarian intervention. In simple yet compelling terms, Lo frames Xunzi's perspective as follows: "A government that practiced *ren* and *yi* would be obligated to use military force to defend the defenseless at times" (Lo 2015a, p. 8). The question is, of course, *at which times*? Or, in other words, under what conditions?

Although seeming to endorse Twiss and Chan's account, Lo confirms a number of claims that ground the interpretation I have offered. Indeed, repeating some of Lo's statements comes close to providing a summary of my arguments: Lo writes, "[I]n Confucianism the greater emphasis is on whether the person who starts a justified war is virtuous, rather than whether any specific kind of action is justifiable" (Lo 2015a, p. 11). More specifically, "Only when power is under the guidance of full virtue can there be a rightful authority to declare war" (Lo 2015a, p. 9). And so, "[B]oth Mencius and Xunzi argued that the rightful authority to declare a morally justified war belonged to the virtuous True King." (Lo 2015a, p. 9; see also p. 22). "A True King would win wars easily and have virtually no enemies" (Lo 2015a, p. 10). Mencius maintained that "a person of *ren* is invincible," and, Lo continues, Xunzi claimed "that the army of benevolence and justice (*renyi zhi bing*) wins a battle without shedding blood. It is not what one does but who one is that ultimately matters the most" (Lo 2015a, p. 11). Lo further suggests that, like the Chinese military strategists, early Confucians maintained that "the best way to employ the military is to subdue the enemy without a fight" (Lo 2015a, p. 20). He continues, "This is to be attained, for the Confucians, by the soft power of moral suasion."

Lo argues that for both Mencius and Xunzi the key to consistent success in using the military resides in being a True King, one who practiced *ren* (humane/benevolent) governance. Such a person would be invincible and, indeed, would even "triumph without actual fighting!" (Lo 2015a, p. 10).[20]

Lo here offers what he considers a "charitable" interpretation, according to which the Confucian strategy does involve a threat of "hard power," but hinges critically on the "soft power" associated with moral credibility. This is not so different from the position I've defended. Although Lo finds Mencius and Xunzi's language to be "exaggerated," he acknowledges that "both Mencius and Xunzi claimed that as soon as military confrontation broke out, the battle would be over before any blood had been spilled" (Lo 2015a, p. 10).[21] Only if one assumes that Mencius and Xunzi did not really mean what they said, can one conclude that Xunzi and Mencius would be relatively open to the use of humanitarian intervention in today's world.

It may seem to us wholly implausible that battles were won bloodlessly, or with only minimal bloodshed. And we may be reluctant to attribute to great thinkers what seems to us to be so implausible. But why then did they state what they did? First, there is at least some truth in it. Benevolent governance really does involve positive feedback loops that make a state strong and secure. And tyranny does the opposite. Moreover, Mencius and Xunzi had practical goals: they were trying to redirect rulers from focusing on the military to focusing on improving themselves and their own governance. For that is the surest way to bring benefit to people.

In assessing the reasonableness of the argument against humanitarian war, it may be useful to remember that even if, occasionally, a war fought for humanitarian purposes will turn out to produce the best consequences overall, that doesn't mean that supporting wars *that seem like they might have that potential* would produce the best circumstances. For we would need to factor in all the times recourse to war to solve problems only created worse problems. It may be better not to be tempted to think we can solve problems with violence even if sometimes we might succeed. In other words, it might be wise to appropriate Elizabeth Anscombe's remark: "But of course the strictness of the prohibition has as its point *that you are not to be tempted by fear or hope of consequences*" (Anscombe 1958, emphasis in original).

FINAL REFLECTIONS

Let me add here that there is a special responsibility for academics and intellectuals in particular to prevent over-zealous leaders from getting into moral trouble, as well as sorting out truth from spin and fabrication. Confucius suggests, in *Analects* 16.1, that rulers are like rhinos or tigers. If one escapes from its cage and causes trouble, Confucius asks rhetorically, "Who is to blame?"[22] The answer, of course, is not the rhino or tiger (the leader), but those responsible for restraining them (scholar-officials). Those who seek power tend to be less than morally perfect, and too often will act from ulterior motives. Thus, well-intentioned intellectuals ought not provide loopholes through which those in power can drive tanks. In the end, in my view, military intervention, even in the case of politically generated famine, would not be justified by Confucian reasoning unless nearly impossible-to-meet conditions were also somehow met, such that the intervention would involve, in Xunzi's words, "punishments, but no battles" (*Xunzi* 15/71/17; 15.1f).

We cannot know exactly what was in the mind of these early Confucians. And it may well be, as it seems, that they really did believe their rhetoric about the ancient sages, and their talk of punishments without battles. But contemporary Confucians need not accept fanciful tales. Further, the early Confucians did not seem to believe that the fabulous feats of the past could be repeated in their own time. Similarly, contemporary Confucians ought to recognize that the criterion for legitimate military intervention—overwhelming support on all sides for the intervention, such that little actual fighting would ensue—is exceedingly hard to meet in our time as well. So, given the absence of leaders uniformly regarded as virtuous, the nature and dynamics of war, and the resistance that intervention inevitably faces in real circumstances, it turns out that offensive war, including humanitarian intervention, is not realistically justifiable by Confucian standards.

A set of issues that have been occasionally alluded to above, but have not thus far been elaborated on, relate to *disingenuous* arguments for war: those involving provocations and propaganda, apparently deliberate distortion of the truth grounded, presumably, in ulterior motives. There is a long history of this, and when evaluating the justification of any particular proposed military action, those precedents should not be forgotten. A handful of examples are given in the following chapter, along with lessons from Confucius's response to an analogous situation.

I also emphasize the importance of remonstrating with wayward leaders. I connect this with Xunzi's theory of natural human dispositions—that people are by nature selfish, and prominent people should not be assumed to be exceptional in that regard. In addition, I suggest that when international tensions are blamed on cultural differences, we are well advised to look first for differences in understandings of the relevant facts on the ground, perhaps supported by differing regimes of propaganda, rather than suspect that the problem stems from irreconcilable differences in cultural values or that it portends an inevitable clash of civilizations.

NOTES

1. As an example, Twiss and Chan cite Yan's tyranny as providing a "just cause" for military intervention (2015a, p. 107). However, Mencius did not approve of Qi's military intervention in this case because the ruler of Qi did not meet the "legitimate authority" criterion (see *Mencius* 2B8; as well as pp. 63–64 and p. 99).

2. Twiss and Chan use emotive language in a one-sided way, highlighting the evils war is intended to end while glossing over the evils that war will inevitably involve. Richard Jackson suggests that this is common in just war theory: "[I]t can be argued that JWT, particularly in terms of its abstracted and decontextualized mode of theorizing, functions to obscure and legitimize the intentional, organized destruction of human bodies" (Jackson 2020, p. 50). It also obscures the psychological trauma on that side of the ledger, namely, that "physical violence, even when undertaken in pursuit of ostensibly just reasons, is a brutal, incomprehensible, traumatic, world shattering experience that is devoid of meaning in its material experience" (Jackson 2020, p. 50).

3. In the introductory section of the book in which Twiss and Chan's chapters appear, Ping-Cheung Lo describes one of Twiss and Chan's chapters, namely "Classical Confucianism, Punitive Expeditions, and Humanitarian Intervention," as arguing that "the Confucian punitive expedition aligns quite closely with the emerging 'responsibility to protect' model in Western discussions" (Lo 2015a, p. 9).

4. Twiss and Chan also mention *Mencius* 7B4 in this context, quoting: "To wage a punitive war is to rectify" (Twiss and Chan 2015b, p. 99, cf. p. 125). Like Bell, they omit what comes next: "If each desires to set themselves straight, what need is there for war (*zhan* 戰)?" See pp. 68 and 98.

5. The word translated simply as "you" here is *wang* 王. It is used as a polite way to address the king: "Your Majesty." But this is the same word that is used for a "true king." To some degree, Mencius may be trading on this ambiguity to suggest that what he says applies to a true king—and so applies to King Xuan only if he takes the proper steps to become one.

6. Julia Ching notes: "The Confucian classic, the Book of History (or Historical Documents), records that King Wu, founder of the Zhou dynasty, ceased military affairs after his conquest of Shang" (Ching 2004, p. 257). In this context, she also points to a passage in the Book of Poetry, which seems to praise King Wu for storing away the weapons of war.

7. I thank Meredith Cargill for allerting me to this point, and the dynamics it implies.

8. Twiss and Chan also cite *Mencius* 1B4, 1B6, 3B5, and 5B9, but only 6B7 is quoted.

9. An analogous point can be made about Qi's invasion of Yan, discussed above. Kim notes, "[A] close reading of the text reveals that the reason Mencius thought Yan deserved a punitive expedition has more to do with the illegitimate royal transmission between Zikuai and Zizhi that did not involve the Mandate of Heaven than with the suffering of the people of Yan itself" (Kim 2017, p. 201; see *Mencius* 2B8).

10. Twiss and Chan acknowledge that appeal to the authority of a true king does not make sense in the modern context (see 2015b, p. 127).

11. For further explanation, see Kim 2017, p. 196.

12. Twiss and Chan also cite *Xunzi* 11.3 and "Book 9 generally," as well as *Mencius* 5A5.

13. The relevant part of the passage, on Knoblock's translation, reads: "Duke Huan of Qi, Duke Wen of Jin, King Zhuang of Chu, King Helü of Wu, and King Goujian of Yue all had harmonious and coordinated armies, so they may be said to have 'entered the precincts.' Nonetheless, since they never possessed the fundamental principles and guiding norms, they thus could become only lord-protector and not kings. Such are the signs indicative of strength and weakness" (15.1d).

14. Note that three of Bell's conditions (see p. 59–60, above) derive from the fourth. That is, the humane intention, the people's welcome, and the support of the world, all derive from the exceptional virtue of intervening agent. As Kim notes, "[B]ecause the agent is virtuous, he has the right motive, people welcome his intervention spontaneously, and the world supports his action" (Kim 2017, p. 199). This applies to my treatment as well. But it is not something I find problematic.

15. Twiss and Chan find the criterion of "potential virtue" to be vague, and, indeed, "vacuous," especially from a Mencian view, in which everyone is "supposed to contain the seeds of virtue within them" (Twiss and Chan 2015a, p. 111).

16. Twiss and Chan admit that these requirements "may well make a great deal of sense," given the potential for the abuse of power. They wonder, however, "How representative must the consent or request of the abused population be? How many states must give prior approval to intervention?" (Twiss and Chan 2015b, p. 129) These are, indeed, difficult questions. I have argued that Mencius and Xunzi suggest that the support must be strong enough to make genuine war unnecessary.

17. Twiss and Chan twice suggest, with apparent skepticism, that it is "unclear whether the Confucians texts were interested in conducting [*post facto* assessments]" of the justification of punitive expeditions (Twiss and Chan 2015b, p. 128, cf. 129). Yet Mencius seems to be clearly providing such a *post facto* assessment when he remarks, "During the Spring and Autumn Period there were no appropriate (*yi* 義) wars" (*Mencius* 7B2). Context indicates clearly that he is referring to punitive expeditions (see extended quotation on p. 58). And relevant *post facto* assessments and criticisms are commonly implied in Confucian texts.

18. See *From Warism to Pacifism: A Moral Continuum* (Cady 2010) for a discussion of various versions of pacifism.

19. For example, see *Mencius* 1A7 (quoted on pp. 69–70), *Xunzi* 9.4 (quoted on p. 116), and *Analects* 16.1 (see p. 122).

20. Lo cites *Mencius* 1A5, 2B1, 4A7, 7B3, 7B4 and *Xunzi* 9.9, 11.8, 15.2.

21. Lo cites *Mencius* 7B3 and *Xunzi* 15.2.

22. For further analysis of *Analects* 16.1 see pp. 121–122, below.

Chapter Seven

From Human Nature to the Clash of Civilizations

Drawing insights from Xunzi's account of natural human dispositions, this chapter explores the duty of intellectuals (and nowadays journalists) to challenge dubious and self-serving[1] narratives. Then I provide several examples of such narratives, each of which justify violence based on a misconstrual of reality and draw some lessons from Confucius's own response to a case of dubious justification for violence. Wars, in particular, are commonly influenced by particular, often dubious, interpretations of events. Finally, I argue that the potential for a "clash of civilizations" is not so much based on cultural differences, or differences in moral worldviews, but rather on differences in interpretation regarding the facts on the ground, especially in cases resembling the examples given. And thus it is important to critically interrogate the purported facts asserted in support of the use of violence, and to remonstrate, or dissent, when these "facts" are found to be dubious.

HUMAN NATURE AND ITS POLITICAL IMPLICATIONS—LESSONS FROM XUNZI

In this section, beginning with Xunzi's very plausible conception of human nature—that people are naturally selfish—I support Xunzi's view that this gives us reason to favor a system of mechanisms to encourage moral development. I further suggest that it gives us reason to be skeptical of arguments supportive of violence, for there may well be selfish motives behind them. Further, when confronted with such situations, intellectuals (and now journalists) have a responsibility to engage in what Confucians would refer to as "remonstrance," that is, to dissent, oppose, and protest against courses of action that they deem inappropriate or inhumane.

Xunzi regards humans as born with a tendency toward selfishness. He may overemphasize this—for I think Mencius was right that people *also* have an innate tendency toward goodness. But Xunzi has a valid point: even if we also have some natural altruistic impulses, we clearly *do* have strong natural dispositions toward selfish behavior too. It does seem true that, as Xunzi himself stresses, our ugly side can be successfully counterbalanced through effort and intelligence. Still, this ugly side *is* there, and it is dangerous to pretend otherwise. This is, arguably, what drove Xunzi to state his position in such a stark and dramatic way: *Xinge*!—natural dispositions are detestable! Artificial constructs (*wei* 偽), including norms of ritual propriety, were invented to help us avoid the negative consequences of uncritically following our selfish dispositions, on the one hand, and to "nurture people's desires and provide for their satisfaction" (*Xunzi* 19/90/3; 19.1a), on the other.

When we consider Xunzi's reasoning, apart from his dramatic slogan about the depravity of human dispositions, it seems that Xunzi is making an important point. And reflecting on his philosophy reminds us that it is critically important to devise various systems to address the fact that people are prone to problematic selfishness. That is, human beings have natural desires that tend to lead them into mischief, which tends to result in conflict that is harmful for everybody. So, especially when developing political systems, we need to devise institutions with *this* truth in mind, even if it is only *half* of the truth. The structures that Xunzi regards as most important are quasi-moral structures that facilitate virtuous self-cultivation and ensure that those who become leaders have achieved a high level of such cultivation. Such people will still have petty desires, like all of us. They still desire food when hungry, and various enjoyments. But they have developed new motivations and aspirations as well. Cultivated people want not only to be honored, but even more to be truly honorable, which is, after all, the surest way to end up being honored (see *Xunzi* 18.9; and Hagen 2011b, pp. 66-67).

What are the implications of all this for us, in our time? First, we must acknowledge the problem in its most dangerous yet often unrecognized form. It is not hard to find examples of depravity, so long as those examples are not too close to home—the Nazis, the Russians, and people like Saddam Hussein, for example, are easily vilified. But there is a strong psychological tendency to deny any truth that would make us feel morally uncomfortable. Yet considering what is near at hand is most important.[2] Acknowledging the selfish aspect of our nature helps us see why we need structures to limit the ability to use violence, even if we can justify it, or rationalize it, in humanitarian terms. And it also may help us to recognize hidden agendas. For example, it may have been more difficult to make the case for the disastrous war in Iraq (not to mention Vietnam) if Americans were not so blinded by exceptionalism that they were easily persuaded that the war was motivated by moral considerations, as well as legitimate concerns for self-defense. But now we know that the Bush administration's arguments were not simply poor ones, based on premises that "turned out" to be false, but were rather *disingenuous*, for the falsity of those premises was known all along (see below)—although this truth is still regularly denied. One implication of *xing e* (that we all have selfish dispositions) is that social mechanisms (*wei*) are needed to prevent the kind of abuses that can be expected from flawed people—that is, all of us, even those who are prominent.

Several examples can be given. First, it is important to have strong legal safeguards that protect individuals from unfair treatment and abuse. This is the work that "rights" does well, and modern Confucians would do well to incorporate such protections one way or another. Second, norms and laws that make wars very difficult to start are also important. Third, we also have to find ways to facilitate moral self-improvement, that is, "learning" in the Confucian sense. And Xunzi's philosophy, in particular, is designed to do just that, by encouraging the cultivation of exemplary dispositions through habitually observing norms of ritual propriety. Because people are selfish by nature, Xunzi reasons, "they must have a teacher and a model before they can become upright" (*Xunzi* 23/113/9; 23.1b). Exemplary moral leadership becomes a precondition for future moral leadership.

Finally, there is a further implication of the inescapabilty[3] of human selfishness, and the difficulty of developing a moral character capable of holding particularly alluring desires in check. Namely, if heads of state are in any sense atypical, it is that they are *ambitious enough to have secured a position of power*, and as a result of that power *they face unusual temptations*. Thus, we ought to be particularly concerned about the morality of world leaders, *including our own*.

Now, what would Xunzi advise? I will mention three things. First, we must notice that atrocities are liable to happen when there is no system to assure the moral cultivation of those in leadership positions. And for moral development, rules do not suffice. Robust norms of propriety designed to facilitate this development are necessary to adequately inculcate virtues. Second, we need to continually engage in what Xunzi calls "dissolving beguilements" (*jiebi* 解蔽). For example, Americans are beguiled by exceptionalism—the view that America is different and better than other countries, that it is an unproblematic force for good in the world. According to this myth, a stronger America means a better and safer world, not just for Americans, but for everybody. This thought blinds Americans to the most severe crimes that their leaders commit or cynically allow. Here *xinge* and *jiebi* go hand in hand. We are blind to the capacity of our leaders to act with cruel selfishness. So, we need to remove our blinders, recognize that people, even our own leaders, without cultivation, can be complicit in atrocities. Xunzi warns that beguilement leads to downfall. He writes, "[The tyrant] Jie died on Mount Li, and [the tyrant] Zhou's head was hung on a red pennant. They could not foresee these events themselves, and there was no one who would remonstrate with them. These are the misfortunes of being obstructed by beguilements" (*Xunzi* 21/102/17; 21.2). Indeed, it is a pattern well established in history that it is the self-deluded over-reaching of empires that leads to their collapse.

The flip side of this is that enlightened government—informed by truth and guided by a sincerely benevolent attitude toward all—results in not only benefits for others but also contributes to a kind of self-realization, while at the same time being the most reliable strategy for the fulfillment of one's desires. As Xunzi writes, "If one becomes a Yao or a Yu [that is, a sage], then one normally is content and at ease" (*Xunzi* 4/15/12; 4.9). Elsewhere he writes, "Everyone desires (*yu*) comforts and honors, and detests danger and disgrace. Only the exemplary person is able to get what he cherishes (*hao*)" (*Xunzi* 8/34/10; 8.11). There are many other passages that also express this idea. Here is just one more:

> If a ruler of the people desires ease and stability, there is nothing like fair policies and loving the people. If he desires glory, there is nothing like exalting ritual propriety, and respecting scholar officials. If he desires achievements and fame, there is nothing like esteeming the virtuous, and employing the able in government. These are the crucial points of a ruler. If these three points are properly dealt with, then all remaining matters will be properly dealt with. (*Xunzi* 9/36/10; 9.4; cf. 21.2)

The third lesson we can glean from Xunzi's philosophy relates to the need for the dispelling of beguilement. It is this: would-be exemplary persons—and I'm thinking particularly of scholars as well as officials (and nowadays journalists)—need to step up and demonstrate their capacity for *ren*—their ability to act unselfishly, to do their utmost in the service of appropriateness (*yi*), out of actually caring[4] (*ai*) for their fellowmen-and-women.

One kind of unselfishness that is particularly needed now is remonstrance, especially on the part of intellectuals. Xunzi writes, "When a ruler is involved in schemes and affairs which go too far, and one fears they will endanger the state, high officials and senior advisors are able to approach and speak to the ruler. Approving when one's advice is used and leaving when it is not is called 'remonstrance'" (*Xunzi* 13/64/1; 13.2). But, alas, few will stand up and say what is true for fear of falling out of favor with those who have influence over the fate of their individual careers.

PROPAGANDA AND THE JUSTIFICATION OF VIOLENCE: FOUR EXAMPLES

At this point I will give four examples in which inappropriate violence has been supported by propaganda. The resulting carnage could probably have been avoided if it had been adequately challenged by intellectuals willing to remonstrate, or dissent. Then I will draw some lessons from Confucius, and finally reflect on the "Clash of Civilizations," arguing that culturally influenced differences in moral philosophies are not primarily what accounts for international differences in moral opinion. Rather, what matters most is differences in propaganda-influenced *interpretations* of "the facts on the ground."

Example 1: The Mukden Incident (known in Japan as "the Manchurian Incident")

When the Japanese describe the beginning of World War II, they explain it this way: "The Manchurian Incident occurred, and Japan was plunged into war."[5] The Manchurian Incident involved an explosion on a railway operated by the Japanese in Manchuria. Of course, at the time, the Japanese propaganda machine did not leave it so ambiguous. Accusing the Chinese of treachery, the Japanese used the incident as a justification to attack a Chinese outpost, and eventually invade China. The Japanese had a "moral argument," but when the true nature of the event is made clear (the Japanese caused the explosion as a deliberate pretext for war),[6] it is apparent that the "moral argument" was a sham and was merely being used in the service of an ambitious power play—a power play that would be, and this is my main point, repudiated from the perspective of *any* moral tradition, and was in violation of the norms of *every* culture.

The Japanese people supported the war for a number of reasons. The propaganda included the idea that what was being created was a "Great East-Asian Co-prosperity Sphere," which suggested that everyone in the region would be benefited. But this of course was also a fiction, a purposefully crafted myth.

Example 2: The 1999 Russian Bombings

In 1999 a series of apartment bombings occurred in and around Moscow. More than three hundred people were killed. These incidents were blamed on "Chechen rebels" and used to justify an attack on Chechnya. The last of these incidents was very interesting. It was a botched job—of one kind or another. A resident in the town of Ryazan contacted the local police after noticing suspicious activity. Namely, he saw three individuals carrying sacks into an apartment building and noted that the license plate of their car seemed to have been altered. When the police investigated, they found a bomb in the basement of the apartment building containing an explosive material called hexogen (also called RDX), and a detonator. The official story at this time, all the way up to Putin himself (who was Prime Minister, but would very soon become President), was that this was another bombing attempt by the Chechen rebels, thwarted by alert residents and local police.

Unfortunately for that story line, the local authorities were not done yet. Two days after the incident, a tip led to the identification of the three individuals, who turned out to be FSB agents—members of the Russian intelligence service. Well, what is going on here? Did the Russians run a covert operation against their own people in order to whip up support for a war? No. Of course not! At that moment[7] the official story changed completely: It was just a drill, they say, after two days of saying that it had been a real attempted terrorist attack. The authorities assured the people that there was never any danger, because the "bomb" was a fake, and the "hexogen" was really just sugar, even though the material had already been analyzed and been determined to contain hexogen. To prove it was really just sugar they actually went so far as to blow it up. The truly amazing part is that all of this was exposed on a live television show[8]—and yet they still got away with it! Putin's popularity soared as the Russian army took "revenge" against Chechnya.

The implicit argument for war, which made it palatable to the Russian people, it seems, was once again based on a false premise. Once the myth is dispelled, nobody, from any cultural tradition, or based on any moral theory, could seriously argue that the Russian bombings (since the Ryazan incident renders all of them suspect) truly justified war. So, the critical thing here again is not any genuine difference over something morally ambiguous, but a failure to dispel the myth.[9]

Now let us consider cases closer to home (for Americans). For it is easy (for Americans) to believe that the Russians, or the Japanese, might be responsible for nefarious deeds. But surely America is different. America wouldn't use deceit to garner support for war. It is this exceptionalism that must be questioned.

Example 3: The Attack on Iraq

According to a report by the Center for Public Integrity:

> President George W. Bush and seven of his administration's top officials, including Vice President Dick Cheney, National Security Adviser Condoleezza Rice, and Defense Secretary Donald Rumsfeld, made at least 935 false

statements in the two years following September 11, 2001, about the national security threat posed by Saddam Hussein's Iraq. Nearly five years after the U.S. invasion of Iraq, an exhaustive examination of the record shows that the statements were part of an orchestrated campaign that effectively galvanized public opinion and, in the process, led the nation to war under decidedly false pretenses.... In addition to their patently false pronouncements, Bush and these seven top officials also made hundreds of other statements in the two years after 9/11 in which they implied that Iraq had weapons of mass destruction or links to Al Qaeda. (Lewis and Reading-Smith 2008)

As Kathryn Olmsted, a history professor at U.C. Davis, notes: "This was a real conspiracy: a conspiracy to perpetrate a fraud on the American public by lying about the intelligence for war" (Olmsted 2009, p. 220).[10]

We could have a significant moral disagreement about whether or not the invasion of Iraq was appropriate only so long as the real intentions of the Bush administration remain unclear. Indeed, the real intentions are still not fully understood, but there can be little doubt that they had more to do with oil and geo-politics than they did with helping the Iraqi people, or protecting the U.S. from weapons of mass destruction, of which UN weapons inspectors *at the time* could find no credible evidence.[11] If we understood the reality of the situation, the war was quite *obviously* immoral[12]—by *anyone's* standards. The conclusion does not need to be based on any kind of nuanced understanding of just war theory, or utilitarian ethics, or Confucian ethics, or any other ethical theory. Roughly speaking, those who agree on the facts tend to agree on the moral conclusion, regardless of their moral philosophy, or political ideology. Of course, this is not to be taken as an absolute rule. We all know that different theories provide different answers in contrived cases. I am merely suggesting that, ordinarily, most of the moral *leverage* is to be found not in moral theory, or differing cultural values, but in the understanding of the *true* circumstances in question.

In using the word "true," and "real," I *do* intend to suggest a reality/appearance distinction, but not of any deep metaphysical sort. I merely mean to suggest that often what *is* the case is significantly different from what *seems* to be the case based on how the issue is framed and presented by the media and by government officials. This is presumably truer in some times and places, and under some kinds of political structures, than in others, but it is a problem that exists even in so-called liberal democracies, and in a *major* way—as the mainstream American press's drumbeat leading into and cheering on the Iraq war attests.

Much of the moral discourse about the war was not just false, but disingenuous. For example, the notions of "preemptive strike" (striking first to preempt an *immanent* attack) and "preventive attack" (attacking so as to prevent another country from becoming a real threat in the future) were routinely conflated—and by people who surely could see the distinction. But even when the moral debate was honest, it was a serious distraction. The muddy waters of moral theory provide a hospitable environment for piranhas; advocates of war can use moral ambiguities to appear as though they have serious arguments. If the focus had been on factual claims, that is, on whether or not the claims that were used to frighten the American people were actually true, the war may have been averted, for by this means its moral baselessness may have been more quickly exposed.

Of course, the war was not *absolutely* baseless, just *morally* baseless. Actions are done for reasons, but *to this day* we have never been given a justification for the war that stands the slightest scrutiny. The true motives, then, are presumably a combination of corporate and imperial interests. And we cannot expect either the government or the corporate media to fess up to those. This makes the need for academics to step up on these issues all the more pressing.

People who believe, as I certainly do, that the invasion of Iraq had more to do with oil and empire than self-defense or altruism, generally oppose the war, whether they are Muslims, liberals, or Confucians, and whether they espouse capitalist or communist economic ideologies. In general, what makes the difference in any truly honest assessment of the morality of the Iraq war is one's view of the reality on the ground, not one's moral or political ideology. And for those who engage in willful deception, we have little reason to believe that their professed ideology is sincerely held. For them, it would seem, morality does not matter. Or else they may have a "higher" morality that, if exposed, most of us would probably call "sick." This is why, perhaps, they resort to deceptive myth making.

Example 4: Torture

Incredibly, whether or not torture is acceptable has become controversial in America, post 9/11. This is only possible because reality is camouflaged by propaganda. That propaganda is premised on the notion that, after 9/11, there was a legitimate fear that there would be further attacks, and that torturing known terrorists was critical to getting information necessary to prevent another attack, and thereby save countless innocent (American) lives. If so, it is argued, then—legal or not—torture was morally justified. In fact, not to torture in such a situation, the logic of this argument suggests, would be irresponsible. *Theoretically* it seems that there is a challenging argument here. But when we attend to the *reality of the situation*, we find that there is no *practical* significance to it, for the whole argument is built on a counterfactual.[13] It assumes that the reason for the torture was to get *information*. However, torture is not really to get *information*,[14] but rather to get (often fake) *corroboration*. It is well established that torture does not produce *reliable* information. Nevertheless, the myth that it might—after all, how can you be *sure*, in a particular case, that it won't?—is necessary for its real purpose to be served.

Why do I suggest that fake corroboration, not real information, is the true purpose of torture? Well, even the Western mainstream media will admit that it is, when it applies to *other people's* torture. For example, on the popular and respected news program, *60 Minutes*, the case of an Iranian man who was tortured in an Iranian prison was covered. The report was explicit and confident in asserting that the reason he was tortured was not so that he would answer their questions truthfully, but rather so that he would say, untruthfully, what Iranian authorities wanted to hear.[15] So that he would, to put it concisely, corroborate their propaganda.[16] Of course there was no suggestion in this context that *American* torture might be done for the very same reasons. This is a form of American exceptionalism. But in fact, we now have good reasons to believe that American torture was done for the same reason. Post 9/11 torture appears designed to acquire corroboration, genuine or not, that would support military action against Iraq. People were tortured, not to stop an attack, but to provide "evidence" for a (non-existent) link between Iraq and al-Qaeda.[17] Just as in Iran, the purpose of American torture was to "corroborate" a line of propaganda.

If people knew that torture was being conducted, not for the purpose of extracting information that would save lives, but rather for building a phony connection between Iraq and 9/11, there could be no argument about whether or not it was justified. (Incidentally, much of the 9/11 Commission Report was based on "information" from people who we now know were subject to "harsh interrogations," including waterboarding.[18])

WHAT WOULD CONFUCIUS SAY?

Although I have focused on relatively recent examples, two of which involved the United States,[19] the dynamics underlying these examples are both universal and ancient. Consider the following example from the *Analects*.

Passage 16.1 of the *Analects* begins with two of Confucius's disciples, Ran Qiu and Zilu, informing Confucius that the Ji clan, whom they serve as ministers, is about to attack a vassal state within the state of Lu. Confucius is outraged and rebukes Ran Qiu for not taking steps to prevent this. Ran Qiu at first says that he and Zilu are against the idea, but then argues, "If we don't seize it now, it will surely be a source of trouble for future generations." Confucius chastises him for his obvious duplicity, and then explains:

> I have heard that [true] leaders of states or clans:
> they do not worry about poverty, but inequity;
> they do not worry about [insufficient] population, but insecurity.
> For if there is equity, there will be no poverty.
> If the population is harmonious (*he* 和) they will not be few in number.
> And if they are secure, there will be no instability.

The situation presented in this passage is analogous to the cases we have just considered, for the reality of the situation is not what Ran Qiu and Zilu presented. The Ji clan is not really worried about a significant threat, but rather is acting on a desire for power. Further, if they were worried about a threat, violence would neither be an appropriate nor an effective way to handle it. Instead, from a Confucian perspective, the leaders should concentrate on improving themselves. But perhaps worst of all, Confucius implies, the whole thing is not so much the fault of the leaders of the Ji clan, but of their advisors, who are Confucius's own disciples, Ran Qui and Zilu. Confucius asks, "If a tiger or rhinoceros gets out of his cage . . . who is at fault?" Obviously, it is not the tiger or rhinoceros that are at fault; the fault lies with those who are supposed to be looking after the tiger and rhinoceros. Likewise, ministers are responsible for safeguarding the ruler with good advice, and yet here they are caught rationalizing folly for the sake of petty advantage. If they could just see through it, and be honest, they could help avert the impending crisis (and apparently it actually was averted).[20] In our contemporary situation, who is going to tell the uncomfortable truth? Who is going to help avert conflict, when even intellectuals are caught up in rationalization and myth perpetuation, setting the stage for a "clash of civilizations"?

Three lessons can be drawn from this passage that are all relevant to the contemporary problem suggested by the four examples. First, rulers are like caged rhinos—they need responsible keepers. It is the responsibility of advisors, scholars and officials, to protect the rulers from their own tendency toward mischief. Second, in order to perform such a function, intellectuals, in particular, and other prominent persons (and journalists too) must see through the thinly veiled motivations of those who would stir conflict, or advocate violence.[21] Challenging though it may be, drawing on rich cultural and philosophical resources, with attention to history, combined with insight into human nature—and Xunzi can be helpful here[22]—it is possible to dispel divisive myths and find ways to live in peace and relative harmony. And third, we are called to exhibit the virtues of *ren* 仁 and *yi* 義 (doing our utmost in the service of others,[23] guided by a sense of appropriateness), and to encourage our leaders to fulfill their moral potential as well. In other words, we need to reject greed, and instead pursue what is, when we attend honestly to reality, clearly the proper course (*dao* 道). Confucius encouraged this repeatedly, and in various ways. To just mention one: Confucius said, "To see what is appropriate and not act is to lack courage" (*Analects* 2.24).[24]

A CLASH OF CIVILIZATIONS?

In providing all these examples, one of my central points is that when self-serving myths are dispelled moral dilemmas often vanish with them. In real cases, when the facts are agreed upon, there is rarely serious dispute between East and West regarding what is morally appropriate and what is not. Any tensions that develop between the U.S. and China—and I worry that they will continue to develop—should be understood in the light of this question: Is it more likely that the tension has to do with real and honest disagreements about what is right and proper, based on different values and ideologies? Or, is it not more likely that the real source of tension stems from a morally *indefensible* selfish power play (probably on both sides)? I mean morally indefensible from *any* sensible moral tradition, whether Chinese, Western, Islamic or some other.

It is not cultural differences that separate us; it is dishonest depictions of reality. If we are willing to focus on reality, and shatter divisive myths, we may yet be able to negotiate a harmonious future. But if not, these myths may be successfully used to foment conflict. Shattering the myths will not be easy or comfortable. It will mean acknowledging exploitive policies from which we benefit and admitting that they are indefensible—by *any* sane set of standards. Really coming to terms with reality in this way must include discontinuing those policies.

Hopefully, dispelling exceptionalist myths will help us to see when our own leaders are acting in morally indefensible ways. And when we find that they are, let us remonstrate with them, insisting that we be led with virtue—with honesty, integrity, and nonviolence.

A CASE TO CONSIDER: SEPTEMBER 11

I have suggested that it is crucial to honestly determine "facts on the ground." But this is not easy, especially when surrounded by Straussian myths, "noble lies"—such as the notion that America is an unproblematic force of good in the world—that enable imperial and other pernicious projects, like the Project for a

New American Century (which will be discussed in the next chapter). In order to achieve a stable international harmony, we must expose these myths. Otherwise, greedy propagandists on all sides will use such myths to divide peoples and generate disastrous conflicts that serve the petty interests of a few.[25]

In other words, the problem is at bottom epistemic. If we don't adequately investigate what is really going on in the "great game" of geo-strategy, we will not be able to achieve the understanding necessary to come to mutually beneficial arrangements.

Consider the advice given in the opening section of the Confucian classic *The Great Learning* (*Da Xue*), which says:

> *Only after one has investigated things* is one's understanding adequate. And only after one's understanding is adequate are intentions genuine. Only after one's intentions are genuine can one properly align one's feelings (*xin*) [literally one's heart/mind]. Only after one has properly aligned one's feelings can one cultivate one's personal character. Only after one cultivates one's personal character can one regulate one's family. Only after one regulates one's family can one order the states. And only after one has ordered the states *can there be peace in the world.*[26]

In short, if we want peace in the world, the process starts with investigating things and gaining an adequate understanding of what is really going on. And then, based on this understanding, we ought to begin to make improvements starting with what is most close to home.

The Project for a New American Century (PNAC) was implemented in the first years of this century directly on the heels of the most devastating terrorist attack in history. The official account of that tragedy, largely unquestioned in the academy, has been pivotal in garnering support for, or at least acceptance of, wars,[27] torture,[28] and the erosion of various legal rights (see Fisher 2008). One's understanding of the defining occurrence of this century cannot be *adequate* if one has made no effort to *investigate* significant aspects of this event, and how it was used to promote or justify violence. One cannot expect peace in this century so long as the most significant event of the century, the formative event, the event foreshadowed in the original PNAC document (see p. 132, below), goes uninvestigated by so many scholars and journalists around the world. It is, after all, the largely unquestioned official interpretation of this event that serves as a key premise for war, and torture, as well as universal surveillance. To allow those who desire empire to determine "the facts" is to give the game away. As Orwell famously wrote, "Who controls the past controls the future" (Orwell 2008 [1949], p. 88). Any viable world order, Confucian or otherwise, must be characterized by open, honest, thorough, and independent inquiry.

In the next chapter, I contrast two Confucian models of world order, one of which imagines a politically unified world. I juxtapose both of these models with the Project for a New American Century, which envisions a world dominated by American military might. Drawing on the work of Sungmoon Kim, I argue in favor of a "Mencian international harmony," in which an informal moral hierarchy between largely independent states plays a significant role in peacefully addressing persistent and ever-evolving international issues.

NOTES

1. I mean "self-serving" in a shallow and immoral sense. Such "self-serving" behavior, from a Confucian perspective, does not actually serve one well. The best thing one can do for oneself, truly, is to develop genuine humanity (*ren*) by doing one's utmost in the service of others. Immoral selfish behavior results ultimately in personal disaster, as well as having other negative effects.

2. It is not my place to focus on what is wrong with China—that's a job for Chinese nationals, who are closer to the relevant information, and in a better position to have an influence. My job is to worry about what is wrong with what is closer to home. Not simply to dwell on the negative, but to strive to correct and improve. Still, the first step to making positive change is to acknowledge a problem.

3. Selfish desires are inescapable—when hungry we desire food—but that does not mean we are doomed to pursue our desires unchecked by other considerations and motivations. Cultivating moral character involves developing the ability to resist inappropriate selfish desires, and also developing moral motivations (see Hagen 2011b for an analysis of this aspect of Xunzi's philosophy).

4. To get a sense of just how far America is from this vision, consider that many of America's most elite males actually participate in, or observe, an annual ritual called "The Cremation of Care," at the Bohemian Grove, in California. The ritual involves a mock human sacrifice of the personification of care. The function of this ritual can be disputed, but it looks as though the participants, wearing hooded robes, are so tired of pretending to care all year long that this ritual provides a catharsis, allowing them to continue to pretend to care for another year. On the bright side, at least they seem to appreciate the power of ritual, though perhaps not ritual *propriety*.

5. This is how it was described, for example, on an exhibit in the Tokyo-Edo Museum. In China the incident is known as "the September 18 incident."

6. Examples like this could be multiplied. Even the Nazis figured they needed pretexts to justify their expansionary military adventures. "Operation Himmler," for example, involved Germans staging attacks on border posts designed to implicate the Polish, and used these incidents to justify their invasion of Poland. This was, in a significant sense, the beginning of the European side of World War II, just as the Mukden Incident was the beginning of the Pacific side of the war.

7. Referring to an announcement that stated that "the planting . . . of a dummy explosive device was part of an ongoing *interregional* exercise" (emphasis added),

the Ryazan Regional FSB issued a statement saying: "This announcement came as a surprise to us and appeared at a moment when the department of the FSB [the Ryazan FSB, that is] had identified the residents in Ryazan of those involved in planting the explosive device and was preparing to detain them" (Litvinenko and Felshtinsky 2007, p. 72). FSB Director Nicolai Patrushev ordered that the suspects not be arrested, but the Ryazan FSB proceeded to arrest them anyway, "considerably roughing them up in the process" (p. 74). According to Litvinenko and Felshtinsky, they were detained "until the arrival from Moscow of an officer of the central administration with documents which permitted him to take the FSB operatives . . . back to Moscow with him" (pp. 74-75).

8. On this Donahue-style TV program, the local residents forcefully challenge the official account. The FSB director and an FSB spokesperson try to argue, sometimes comically unconvincingly, that there is an innocent explanation for all of this. And a former FSB director, Evgueni Savostianov, looking incredulous, comments on the claim that it was just a drill: "I don't get it. Why did it take two whole days to tell the world it was an exercise? Frankly it's incomprehensible." Portions of this show can be seen in the documentary *The Assassination of Russia,* which is at this time freely available on video.google.com.

9. For more incriminating details, see Litvinenko and Felshtinsky 2007, especially pp. 54-75. Alexander Litvinenko is the former KGB and FSB spy turned whistle-blower who was assassinated in London in 2006 by being poisoned with polonium-210.

10. Olmsted cites Paul O'Neill, formerly secretary of the treasury in the G. W. Bush administration: "'It was all about finding *a way to do it,*' O'Neill said. 'That was the tone of it. The President saying, "Fine. Go find me a way to do this"'" (Olmsted 2009, p. 219, emphasis in original). Olmsted also cites Colin Powell's former chief of staff, Lawrence Wilkerson, as describing Powell's now-infamous UN speech (widely lauded at the time) as "a hoax on the American people, the international community and the United Nations Security Council," as well as a British memo that reveals that "the intelligence and facts were being fixed around the policy" (Olmsted 2009, p. 220). The policy, of course, was war.

11. As recorded in an article entitled "Blix Insists There Was No Firm Weapons Evidence" (MacAskill 2005), Hans Blix, who led the UN weapons inspectors, asserts: "We reported consistently that we found no weapons of mass destruction." In addition, a memo dated January 31, 2003 (nearly two months before the beginning of the war) confirms that both George Bush and Tony Blair themselves believed that weapons were not going to be found. And for this reason they considered other ways to justify an invasion. Bush even told Blair that the U.S. already had plans "to fly U2 reconnaissance aircraft painted in UN colors over Iraq with fighter cover." The purpose would be to provoke a response by Saddam that they could then use to justify the invasion they so clearly wanted. This is described in an article entitled "Confidential Memo Reveals US Plan to Provoke an Invasion of Iraq" (Doward *et al.* 2009).

12. Naomi Kline explains, "The widespread abuse of prisoners is a virtually foolproof indication that politicians are trying to impose a system—whether political,

religious or economic—that is rejected by large numbers of the people they are ruling" (Kline 2007, p. 156).

13. According to U.S. Senator Sheldon Whitehouse: "[T]here has been a campaign of falsehood about this whole sorry episode. It has disserved the American public. As I said earlier, facing up to the questions of our use of torture is hard enough. It is worse when people are misled and don't know the whole truth and so can't form an informed opinion and instead quarrel over irrelevancies and false premises.... [T]he presumption of truth that executive officials and agencies should ordinarily enjoy is now hard to justify. We have been misled about nearly every aspect of this program. ... [M]easured against the information I have been able to gain access to, the story line we have been led to believe—the story line about waterboarding we have been sold—is false in every one of its dimensions. I ask that my colleagues be patient and be prepared to listen to the evidence when all is said and done before they wrap themselves in that story line" (Whitehouse 2009, pp. S6360-S6361).

14. Klein writes, "As a means of extracting information during interrogations, torture is notoriously unreliable, but as a means of terrorizing and controlling populations, nothing is quite as effective" (Kline 2007, p. 156).

15. "Asked what the purpose of torture in Iran is, [Lily] Mazahery [an Iranian-American attorney] told [Anderson] Cooper, 'To get these prisoners to say things that the regime wants them to say.'" The episode, which aired on April 5, 2009, is entitled, "How Ahmad Batebi Survived Torture In Iran."

16. It is likewise not taboo in America to say that American soldiers were tortured during the Korean War "for the purpose of eliciting *false* confessions for propaganda purposes," to use Senator Carl Levin's words (Levin 2011). Levin also makes the general point about the psychological dynamics of torture. He cites Lieutenant Colonel Morgan Banks, the senior Army SERE psychologist, who, back in 2002, warned: "If individuals are put under enough discomfort, i.e., pain, they will eventually do whatever it takes to stop the pain. This will increase the amount of information they tell the interrogator, but it does not mean the information is accurate. In fact, it usually decreases the reliability of the information because the person will say whatever he believes will stop the pain" (Levin 2011).

17. Lawrence Wilkerson, former Chief of Staff to Secretary of State Colin Powell, writes, "So furious was this effort that on one particular detainee, even when the interrogation team had reported to Cheney's office that their detainee 'was compliant' (meaning the team recommended no more torture), the VP's office ordered them to continue the enhanced methods. The detainee had not revealed any al-Qa'ida-Baghdad contacts yet. This ceased only after Ibn al-Shaykh al-Libi, under waterboarding in Egypt, 'revealed' such contacts. Of course later we learned that al-Libi revealed these [non-existent] contacts only to get the torture to stop" (Wilkerson 2009). The forced, false "information" that al-Libi provided was used by Colin Powell to justify an invasion of Iraq in his notorious speech at the United Nations.

18. According to an article entitled "9/11 Commission Controversy," in the series *Deep Background: NBC Investigates*, "[M]uch of what was reported about the planning and execution of the terror attacks on New York and Washington was derived from the interrogations of high-ranking al-Qaida operatives. Each had been subjected

to 'enhanced interrogation techniques.' Some were even subjected to waterboarding. ... [Philip] Zelikow [the 9/11 Commission executive director] admits that 'quite a bit, if not most' of its information on the 9/11 conspiracy 'did come from the interrogations'" (Windrem and Limjoco 2008).

19. One reason I focused on the United States is precisely because I am an American, living in America. Thus, the United States is what I know best. But also, it is all too natural, easy, and usually self-serving to point the finger at others. Further, there is usually little leverage in doing so. If there is a problem in America, then, as an American, it is *my* problem. In other words, we each have a special duty to correct our own respective countries.

20. Edward Slingerland explains: "As Liu Fenglu notes, there is no record of an attack on Zhuangyu in the *Annals*, which indicates that the attack never occurred—conceivably as a result of the rebuke recorded here dissuading the Ji Family" (Slingerland 2003, p. 191).

21. Confucius elsewhere also suggests that one must always make one's own moral assessments, saying "If the masses detest something, one must be sure to look carefully into the matter; and if the masses are fond of something, again one must be sure to look carefully into it" (*Analects* 15.28). For more on Confucius's position on self-reliance in learning, see *Analects* 5.9, 7.8, 9.8, 9.11, and 15.16.

22. As discussed earlier in this chapter, Xunzi is famous for the slogan *xing e* 性惡 (innate human dispositions are detestable), by which he means that people are unfortunately endowed with a problematic motivational structure. In particular, we are born selfish, and require moral training to become good. See *Xunzi* 19/90/3; 19.1a, quoted on p. 145 n5.

23. Regarding this aspect of the meaning of *ren* 仁, consider the following: Confucius said, "As for *ren* 仁, undergoing difficulties, and only then reaping rewards can be called *ren*" (*Analects* 6.28). His disciple Zengzi says, "Scholar-officials must be strong and determined, for their burden (*ren* 任) is heavy and their way (*dao* 道) is long. They take *ren* 仁 (genuine-humanity) as their own responsibility (*ren* 任). Is this not heavy? And they carry it until their dying day. Is this not long?" (*Analects* 8.7) For further discussion of the concept *ren* 仁, see pp. 4–5.

24. One could also cite the following, the last two of which are Confucius's comments: "The master refrained from four things: speculation, dogmatism, inflexibility, and selfishness" (*Analects* 9.4). "At the sight of profit, those who think of what is appropriate, at the sight of danger, those who would offer their lives, when in want for a long time, those who do not forget the words they lived by in better times, they surely may also be considered accomplished people" (*Analects* 14.12). And, "Inappropriately acquired wealth and status are as relevant to me as the drifting clouds" (*Analects* 7.16).

25. Those tempted to write this off as "conspiracy theory" are hereby directed to the philosophical literature on conspiracy theories, much of which can be found in Coady 2006 and 2007. The articles by David Coady and those by Charles Pigden are particularly compelling. See also Hagen 2020b and *Conspiracy Theories and the Academy: A Philosophical Critique* (Hagen, forthcoming) as well as *Taking Conspiracy Theories Seriously* (Dentith 2018). The bottom line of this work is that, although

there have been many attempts to argue that conspiracy theories, or some definable subset thereof, are somehow not worthy of belief, all such attempts have failed. This finding should not be too surprising, since all sides agree that at least *some* conspiracy theories turn out to be true (see, for example, Sunstein and Vermeule 2009, p. 206). And thus, each conspiracy theory ought to be judged on its own particular evidentiary merits and demerits. This requires considering the evidence.

26. The same passage also expresses essentially the same set of ideas in reverse order: "In antiquity, those who wanted their distinct influential-virtue (*de*) to shine forth in the world first ordered their states. Wishing to order their states, they first proceeded to regulate their families. Wishing to regulate their families, they first proceeded to cultivate their personal-character (*shen* 身). . . . Wishing to make their thoughts and intentions authentic, they first proceeded to extend their understanding. Extending understanding resides in examining things" (*Da Xue*). Notice that the success at the international level traces back to self-cultivation, and also note the ultimate importance of examining things.

27. The "Joint Resolution to Authorize the Use of United States Armed Forces against Iraq" (October 2, 2002) gave the President the power "to take the necessary actions against international terrorists organizations, including those nations, organizations, or persons who planned, authorized, committed, or aided the terrorist attacks that occurred on September 11, 2001" (cited in Ball 2007, p. 12). Regarding propaganda for the Iraq war, see Lewis and Reading-Smith 2008, quoted at length on pp. 118–119, above.

28. Naomi Klein explains: "[A]fter the attacks of September 11, [the Bush Regime] dared to demand the right to torture without shame . . . according to [the] new rules, the U.S. government was free to use the methods it had developed in the 1950s under layers of secrecy and deniability—only now it was out in the open. . . . [T]he White House used the omnipresent sense of peril in the aftermath of 9/11 to dramatically increase the policing, surveillance, detention and war-waging powers of the executive branch—a power grab that the military historian Andrew Bacevich has termed 'a rolling coup'" (Klein 2007, pp. 51-52, and 377).

Chapter Eight

Two Visions of Confucian World Order

The philosopher Bertrand Russell, and many others, have proposed that uniting the world under a single governing structure with a monopoly on military force could rid the world of the scourge of war. In this chapter, I question the wisdom of such a strategy as I evaluate the viability of two visions of Confucian world order. The two Confucian views are labeled "Xunzian global order" and "Mencian international harmony." The former envisions a single world order in a strong sense, in which a world government with benevolent leadership replaces national sovereignty, while the latter envisions a world in which a stable harmony is achieved among distinct nations and relative influence is determined by a moral hierarchy among them. I argue in support of the latter vision.

While it is possible that Confucianism will never achieve the kind of prominence and influence in the world that the musings contained in this chapter assume, it very well might. And it is worth thinking about reasonable contingencies. Further, applying Confucian ideas to conceptions of world order is not only interesting in its own right, but may well provide us with useful insights even if the West remains the dominant shaper of the international or global order. I will sketch and evaluate two Confucian visions of world order that contrast sharply with the Project for a New American Century, which envisioned an empire secured by coercive military force.

ANOTHER IMPERIAL PROJECT?

One year before the attacks of September 11, 2001, a document entitled *Rebuilding America's Defenses* was produced by a neo-conservative think-tank called the Project for a New American Century (PNAC)—it is sometimes referred to as "the PNAC document." It is a blueprint for strengthening U.S. hegemony. Recognizing that their recommendations were a tough sell, the document's authors write, "[T]he process of transformation, even if it brings revolutionary change, is likely to be a long one, absent some catastrophic and catalyzing event—like a new Pearl Harbor" (Donnelly 2000, p. 51).[1] Then, after 9/11, the supporters of this project, including Dick Cheney, Donald Rumsfeld, Paul Wolfowitz and other prominent neo-conservatives, shamelessly exploited the tragedy, and the fear it engendered, in the pursuit of the initiatives outlined in the PNAC document (see Scott 2007, pp. 23-24). As Naomi Klein puts it, "The mantra 'September 11 changed everything' neatly disguised the fact that for free-market ideologues and the corporations whose interests they serve, the only thing that changed was the ease with which they could pursue their ambitious agenda" (Klein 2007, pp. 377-378). The results included the disastrous Afghanistan and Iraq wars, and the swelling of U.S. debt. While the authors of the PNAC document achieved most of their proximate goals (a military buildup, two wars, profits for the properly connected, and a more powerful executive branch[2]), the ultimate results appear to be virtually the opposite of those desired. The American imperial project is imploding. The question is no longer: "Will America achieve full-spectrum dominance?" It is: "How will America regain international credibility?"

Looked at from a Confucian perspective, the root problem was a moral failure, and the result was predictable. As Xunzi wrote, "One who uses the state to establish appropriateness (*yi*) will be a true king . . . those who scheme for power will perish" (*Xunzi* 11/49/15; 11.1a).[3] To be generous, there is a sense in which the PNAC group can be seen as well-meaning utopians. They just had an ineffective strategy. But the reason it was ineffective is that the moral base was a mirage. At best, the neo-conservative authors were self-deluded, for their project was transparently self-serving. The question of this chapter is, assuming there is no escaping the need to reason about what structures and norms should govern international relations: *Is there a viable Confucian alternative strategy to achieving a better world?*

I am not proposing "A Project for a New *Chinese* Century," that is, a Chinese version of the Project for a New American Century. Such a project would ultimately fail for the same reasons that the PNAC endeavor failed. Further, truly consistent commitment to Confucian principles entails both a commitment to diversity, through which harmony is achieved, and to appropriate treatment for all. More specifically, it entails, as a norm, that those higher in various hierarchies do their utmost in the service of those below them. As it applies to the world as a whole, the hierarchy can be thought of either as existing within a unified global order or as existing among largely independent nation states. I will consider Confucian versions of both, extrapolating from the philosophies of Mencius and Xunzi.

First, I will describe "a Xunzian global order." Although I believe that such a truly Confucian global order would be superior to an American empire, and perhaps an improvement upon the instability of today's world, I have serious reservations about *any* global regime. And so, following my description, I will explain why I do not endorse it. Then I will suggest an alternative Confucian vision, a Mencian *international* harmony. For our purposes, the most significant difference between the two Confucian visions is that a Xunzian global order envisions a single world order in a strong sense, whereas a Mencian international harmony envisions a stable harmony among distinct nations,[4] although there would be a certain kind of hierarchy among them. (Near the end of the chapter I will consider an alternative reading of Xunzi, according to which his view is not so different from that of Mencius.)

A XUNZIAN GLOBAL ORDER

One of Xunzi's most central ideas is that original human dispositions are "detestable." That is, people have a tendency to get into conflict with one another due to natural selfish inclinations.[5] And so, by extension, according to Xunzi, the root cause of international conflict is that people are by nature selfish (Yan 2008, p. 154). Xunzi thought that in order to achieve a stable harmony, which is the ultimate Confucian goal, peoples' motivational structure needed to be altered and improved. And Xunzi believed that a well-devised system of ritualized norms of propriety was vital to bringing about the necessary transformation of character. Without such a system and the "good people" that resulted, peace would not be possible. Xunzi writes, "People cannot survive without ritual propriety. Endeavors will not succeed without ritual propriety. Neither the state nor the home will be at peace without ritual propriety" (*Xunzi* 2/5/15; 2.2). Ritual propriety involves making appropriate (*yi*) distinctions (*fen*),[6] and this in turn leads to a social hierarchy designed to prevent conflict. Xunzi writes, "How are we able to put social divisions (*fen*) into practice? I say it is a sense-of-appropriateness (*yi*). If a sense-of-appropriateness is used in forming divisions (*fen*) then there will be harmony" (*Xunzi* 9/39/11; 9.16a).[7] (Harmony presupposes differentiation, and in the Confucian system this differentiation involves hierarchy. However, in a Confucian hierarchy various stations are defined primarily by their responsibilities—with those higher in the hierarchy having more burdensome[8] responsibilities. So, it is at least as much a hierarchy of responsibility as it is one of privilege.)

Ultimately, on Xunzi's account, the hierarchy should lead to a single individual at the top. The family provides a model for this. Xunzi writes, "The ruler is the most exalted in the state. The father is the most exalted in the family. Exalting one [results in] order; exalting two [results in] chaos. From ancient times to the present, there has never been a situation that was able to endure for very long with two exalted, each contending for respect" (*Xunzi* 14/67/17; 14.7; cf. 9.3). Xunzi's line of thought here seems to lead to a unified system with a single morally cultivated individual at its head. Sumner Twiss and Jonathan Chan describe Xunzi's vision this way: "Ideally, the entire political world would be united under the humane rule of a sage-king whose prime duty is to work to benefit all people" (Twiss and Chan 2015a p. 96). This Confucian ideal would be, as Twiss and Chan describe it, "a vital, cooperative, non-violent, and flourishing polity on a grand scale, held together and harmoniously functioning according to elaborate ritual forms" (Twiss and Chan 2015a p. 96). The process leading to this result can be called Xunzian globalization.

The point of unifying the world, in Xunzi's view, was to create a stable harmony. And it would be only "a wise and insightful exemplary person"[9] or "a person of *ren* (humanity),"[10] that would be up to the task. The reason it takes a moral exemplar to do this is that moral persuasion, not force, is the key to unification. As Yan Xuetong explains: "In [Xunzi's] view, . . . the power status of 'having the world' is gained through peoples' and other states' voluntary submission, rather than through use of force" (Yan 2008, p. 148).[11] This is not mere idealism. As a Chinese adviser is reported to have told Kublai Khan: "[O]ne can conquer the empire on horseback, but one cannot govern it on horseback" (Bawden 2020). Indeed, there is considerable historical evidence of the inadequacy of coercive power to maintain control, including the recent difficulties in Iraq and Afghanistan. So, it is at least plausible that virtuous domestic and foreign policy would be comparatively better than less admirable, more forceful alternatives.

This is, of course, not to champion a Huxlian "Brave New World" in which people are cajoled, seduced, entranced, and beguiled by relentless propaganda, infotainment, and distraction, a world in which basic needs are met for the purpose of maintaining pacification—what the Roman's called "bread and circus." No reasonable interpretation of *ren* would allow that.

In the West, we tend to want fixed rules that will protect us, such as a robust Bill of Rights—and I agree that having certain assurances, one way or another, is very important. With the right institutions, thought Immanuel Kant, there can be justice even in nations run by devils.[12] Xunzi would disagree. World peace, he thought, cannot be secured by merely having the right political institutions. Rather, in Xunzi's view, what is most important is not good laws or policies, but *good people*. And it takes ritual propriety, not laws, to cultivate goodness in people.[13] As alluded to in the previous chapter, Xunzi writes, "People's original dispositions are detestable, they must have a teacher and a model before they can become upright" (*Xunzi* 23/113/9; 23.1b).[14] He stresses this over and over, at one point elaborating:

> If there are good persons, then [the political order] will survive. If there are no good persons, then it will be lost. Laws and norms are the starting point of orderly government. Exemplary persons are the wellspring of these laws and norms. Thus, if there are exemplary persons, this is sufficient for a vast state, *even if laws and norms are omitted*. But if there are no exemplary persons, then although a state may be equipped with laws and norms, it will misstep in the application of priorities, and its inability to cope with changing circumstances will suffice to result in chaos. (*Xunzi* 12/57/4; 12.1)[15]

Elsewhere, Xunzi writes:

> When it comes to managing a myriad of changes, adjudicating the myriad phenomena, nurturing and cultivating the myriads of common people, and simultaneously governing the whole empire, those who do their utmost in the service of others (*ren ren*) are the best. Their wisdom and planning are sufficient to manage the changes. Their humanity and generosity are sufficient to pacify the common people. And, the resonance of their virtuous charisma (*de*) is sufficient to transform them. If such people are obtained, there will be order; if not, there will be chaos. (*Xunzi* 10/43/19; 10.5)[16]

And also:

> Exemplary people are vital for uniting the way (*dao*) with law (*fa*). They must not be neglected for even a short time. If they are acquired there will be order; if they are lost there will be chaos. If they are acquired there will be peace and security; if they are lost there will be danger. If they are acquired [the state] will survive; if they are lost it will perish. Thus, there are cases in which there were good laws, and yet there was chaos. But I have never heard of a case, from ancient times to the present, of their being exemplary people and yet chaos. A tradition says, "Order is produced by exemplary people; chaos is produced by inferior people." This expresses my point. (*Xunzi* 14/66/24; 14.2)

It is clear that, in Xunzi's view, we need "good," morally-cultivated people in leadership if we are going to have good government, that is, a government that can be trusted to truly look after the needs and interests of the people and do so effectively. On this issue, Xunzi's view is at least arguably more plausible than Kant's. But how are we to make *sure* we have moral leaders? On the one hand, Xunzi's philosophy is most fundamentally a system designed to produce such people, emphasizing ritual propriety. But on the other hand, even if this is a good system, even if it is better than any other, can it really be considered reliable enough in the context of a world dominated by *global* institutions?

WHAT IS WRONG WITH A GLOBAL WORLD ORDER?

Many people would probably sympathize with the attitude of Bertrand Russell, who, near the end of his book *The Impact of Science on Society* seemed to look favorably on the prospect of a "single government of the whole world, possessing a monopoly of armed force and therefore able to enforce peace" (Russell 1953, p. 113), and also to ration food and materials.[17] However, earlier in the same book Russell makes a remark that should give us pause. But before I tell you what that remark is, I'll briefly outline his argument: (1) Scientific advances generally make despotic rule easier. They facilitate propaganda, surveillance, and other methods of control.[18] (2) It follows that a scientific dictatorship would be *internally* stable. (3) The only real threat to such a system would be war. (4) The tyranny that characterizes scientific dictatorship is the antithesis of freedom, and such a system prevents *scientific* freedom. (5) Since *free* scientific inquiry will be more productive than scientific inquiry in a stifling tyranny, scientific *democracies* will outcompete scientific *dictatorships*, including in military technology. (6) So, democracy will in the long run triumph.

Well, that is all very reassuring. However, Russell then introduces a caveat, the one that should give us pause: Unless the tyranny becomes global![19] Then, the whole world would be stuck with a *stable* tyranny.

So, any desirable world order should be designed with sufficiently reliable mechanisms for preventing the devolution into *that*. This seems extremely important, and it is a task not easily accomplished. Yet, from what I can see, those pining for a "New World Order" show no sign of interest in this issue, or even awareness of the problem. Instead of addressing this issue philosophically—that is to say, with stubborn persistence—the world lunges awkwardly toward larger regional governance with less democracy and fewer safeguards. The formation of the European Union is perhaps the clearest example. And, what had been called the Security and Prosperity Partnership of North America, which involved an almost entirely undemocratic process, represents a less mature stage of what may become a similar unification process. (That is, unless the European Union falls apart first.)

Regarding Xunzian global order, the following remark by P. J. Ivanhoe is revealing: "Xunzi saw [human distinctions] as the unique possibility for universal harmony and flourishing. If one could understand and master the *dao*, things would fall into place on a universal scale. The Way could protect one from all harm, offer one every benefit and bring peace, order, and prosperity to all the world" (Ivanhoe 1991, p. 317). In this view, Xunzi provides a strategy for Good to *win*, once and for all. Alan Watts, a popularizer of Asian philosophies, suggests that while Good must always be *winning*, it must not triumph. He calls the effort to eradicate evil "the game of black *versus* white," and suggests that the notion that white *must win* is paradoxical, because if white does win, the fun is over (see Watts 1966, pp. 25-52). I would add that "white must win" has an ugly history, and present. The project to win once and for all is particularly suspect when "Good" must use morally dubious methods to achieve its supposedly pure ends, destroying cities in the name of peace. "The war to end all wars," of course, didn't. Imposing democracy, enforcing freedom—we see these projects failing in Iraq and Afghanistan, predictably. Consider one more example: In order to eradicate wrong-thinking, America's "pre-eminent legal scholar"[20] and Obama appointee, Cass Sunstein, along with his Harvard Law School colleague, Adrian Vermeule, recommend "cognitive infiltration" of groups whose views they deem to be based on a "crippled epistemology" (see Sunstein and Vermeule 2009)—as if the deceptive practices involved in such infiltrations were epistemically faultless.[21] We should lie, because Truth *must win*. On this view, we can't just muddle through, allowing people to freely assemble and share their ideas trusting that, although not all beliefs will be true, or even well-grounded, in the long run this is the best practice, both epistemically, and in terms of exhibiting appropriate tolerance, if not respect.

The attitude that "Good must win" may in part explain the attraction of what some call a "New World Order." Looked at generously, the desire for a New World Order can be seen as an attempt for Good to finally *win*. (Less generously, it is seen as an attempt for Evil to win—but never mind that.) In other words, it is not a strategy meant to help us muddle through circumstances as they arise, ameliorating problems and working towards modest ends-in-view. It is rather an attempt to *settle* the world into a *fixed*, stable, secure condition. It is an "end-game," to bring an end to risk, and make us secure from all evils. It is utopian—if it is not dystopian. I will not argue here that this will either fail or succeed-disastrously, but I do want to at least suggest that those are among the most likely outcomes. And I want to present an alternative, which I believe can be situated rather comfortably in traditional Chinese ways of thinking. Namely, I would like to suggest that the most attractive Confucian vision is one that involves a continual process of doing one's best to harmonize the constantly changing ingredients of international relations. And this leads to the consideration of a model of Mencian international harmony, as an alternative to Xunzian global order.

MENCIAN INTERNATIONAL HARMONY

In *The Mind of Empire: China's History and Modern Foreign Relations*, Christopher Ford argues, "Confucian thinkers assumed that a perfectly virtuous ruler would naturally come to hold sway over all humankind" (Ford 2010, pp. 35-36). Ford argues that this is partly a consequence of China's particular history (and partly an early philosophical assumption).[22] He writes, "Chinese history provides no precedent for the stable, long-term coexistence of coequal sovereigns, and its traditional ideals of moral governance and statecraft cannot even admit such a possibility" (Ford 2010, p. 5). Indeed, the so-called "Warring States period" is the exception that proves the rule, according to Ford. He explains, "The Warring States period and the Qin/Han unification may in this crucial respect, therefore, be regarded in the Chinese tradition as a *single* model for statecraft—the former being merely the subparadigm of interstate relations undertaken *for the purpose* of achieving the natural and inevitable ideal of unitary governance presented by the latter" (Ford 2010, p. 26). I have no objection to unitary governance as a far-off *ideal*. Something that should happen, and, if the Confucians are right, will happen as a matter of course when all the right conditions are met in an enduring way—which is not in the foreseeable future. In the meantime, a Confucian focus should be on achieving these conditions, which involves independent states striving to be more *ren* (humane, benevolent) than the next.[23] For example, in *Mencius* 1A7, Mencius describes how benevolent government results in security and popularity for the king. It is true that the context suggests that this would naturally, over time, lead to unification of the world under a true king, as Ford stresses. Nevertheless, it also presents a picture of what could be a fairly stable "moral competition" among independent states with porous borders—or at least the possibility of immigration and emigration.

In a compelling article, "Mencius on International Relations and the Morality of War: From the perspective of Confucian *Moralpolitik*," Sungmoon Kim argues that, on Mencius' view, a stable harmony can be best maintained under a "Confucian international moral hierarchy" among states.[24] He describes Mencius' position as follows:

> Mencius . . . believed this seemingly helpless world could be improved and potentially transformed into the moral community of states where both big/strong and small/weak states have a proper, ritual-ordered, moral-political place. In Mencius' view, it is by creating such an international condition that each state can best serve the supreme moral-political goal of the welfare of the people. (Kim 2010, pp. 39-40)

According to Kim:

On the one hand, Mencius was particularly concerned with the political autonomy of small states in the face of the pathological *realpolitik* of war and aggression—a political self-determination that he thought was indispensable to the welfare of the people and could only be attainable under the international moral hierarchy between big/strong and small/weak states. Yet, on the other hand, he clearly understood that despite the mutual reciprocity involved in international moral hierarchy, without the determined moral commitment and responsibility of the big/strong states international peace would be practically impossible. Thus understood, Confucian international *moralpolitik* upholds the political self-determination or autonomy of the (small) state without making it morally and culturally autonomous. What qualifies all states to be legitimate members of the international community of moral hierarchy is the shared understanding and practice of morality in terms of the *wulun*. (Kim 2010, pp. 45-46)

The *wulun*, or "five relations," refers to the relations of parent/child, ruler/minister, husband/wife, elder-sibling/younger-sibling, and friend/friend. But more generally it refers to the normative notion that there are appropriate roles and responsibilities that go with one's relative position in a family, society, or by extension, any social structure. And so, states of various sizes, strengths, and positions in the international hierarchy would have corresponding and mutually-reciprocating responsibilities. For example, the big state should "submit to" (take care of) smaller states, and smaller states would "submit to" (defer to) larger ones (see Kim 2010, pp. 40-41). This is analogous to a parent's responsibility to care for his child, and the child's corresponding responsibility to defer to the parent. Kim cites a passage in which Mencius is asked if there is a way to promote good relations between neighboring states, and provides Mencius' response:

> There is. Only a benevolent man can submit to [i.e. take care of] a state smaller than his own. This accounts for the submission of Tang to Ge and King Wen to the Kun tribes. Only a wise man can submit to [i.e. defer to] a state bigger than his own. This accounts for the submission of Dai Wang to the Xun Yu and Gou Jian to Wu. He who submits to a state smaller than his own delights in Heaven [*tian*]; He who submits to a state bigger than his own is in awe of Heaven. He who delights in Heaven will continue to enjoy the possession of the Kingdom while he who is in awe of Heaven will continue to enjoy the possession of his own state. (*Mencius* 1B3)[25]

The hierarchy involved in this arrangement is a moral one. And, unlike the fixed relations found in a family, a country's position in this hierarchy may change. While size matters, it is more the ability to help that qualifies one for the burdens of "care giving." Sweden, for example, may be among the states that come closest to living up to the Confucian idea of benevolent governance, in terms of taking care of its people, and in terms of foreign aid. In the Confucian worldview it is believed that there are dynamics that play out, ultimately, in a roughly moral direction, and so the influence of benevolent countries will, in the long run, increase. Mencius understands this in relation to *tian*. As Joel Kupperman explains the concept:

> *Tian* . . . refers to a kind of impersonal—but not morally neutral—cosmic order, which manifests itself especially in natural and political disasters that undermine wicked rulers.[26] The assumption is that the universe has as it were its own purposes. Insofar as we are part of the universe, we can be attuned to the purposes of *tian*. Or, like the wicked rulers, we can be deeply opposed to these purposes. . . . [A]t some critical moments . . . there is a tilt against wickedness. (Kupperman 2007, pp. 101–102)

Of course, it would be unwise to pin our hopes for peace on a mere article of faith. But the relevant elements of this view are *not* simply ungrounded assumptions. After all, the notion that there are consistent dynamics in the way the world works, at least some of which trace back to regularities in human interactions based on roughly universal human dispositions and tendencies, can hardly be denied. This is precisely why the failure of the PNAC project was predictable. There is reasonableness to Mencius' *moralpolitik*. People *do* respond favorably to being treated well, and they *do* respond unfavorably to abuse. The logic regarding moral competition is similar to the logic of Russell's argument about scientific competition. Truly humane or benevolent (*ren*) government, which Confucians view as an expression of our fullest humanity,[27] tends to produce favorable conditions that ultimately provide decisive competitive advantages. Even if the language of *"tian"* sounds mystical, it is translatable to a language and logic that is downright commonsensical.

A REVISED INTERPRETATION OF XUNZI'S VISION OF WORLD ORDER

A rather straightforward reading of Xunzi, as described above, suggests that his ideal world order involved a unified political structure for the entire world. This is reflected in the interpretation, for example, of Yan Xuetong, who has maintained that "[Xunzi] regarded unification of the world as true Kingship's highest goal" (Yan 2008, p. 147). Indeed, there are many passages[28] in which Xunzi does seem to suggest this, one of which I will discuss shortly. However, as the original publication of this chapter mentioned in a footnote, "I think that it may be possible to understand Xunzian 'unification' as much less monolithic than assumed [by Yan and others]" (Hagen 2016a, p. 172 n14). Given my original purposes—fleshing out two distinct Confucian conceptions of world order—I chose not to challenge the common and straightforward understanding of "unification." But here, I will briefly develop the alternate understanding of Xunzian "unification" that I had hinted at previously.

A key passage that seems to suggest that Xunzi advocated a unified political order, under a single benevolent ruler, is the following. Xunzi writes, "The person of genuine-humanity (*ren ren* 仁人) uses his state not simply to hang onto what he has and no more, but to unify [all] people" (*Xunzi* 10/48/17; 10.14). In translations of this line, a word like "all" is typically added,[29] and the translation I use adds it as well, in brackets—"to unify [all] people." But "all" is not in the original text. So, *taken by itself*, the sentence could mean that the person of genuine-humanity does not merely want to *hold on to power* in his state, but to *unify* the people *of his state*. In other words, he engenders a cohesive social harmony among those to whom he is primarily responsible.

Now, to be clear, I do not think this is the correct interpretation. But the point is that it is a perfectly reasonable interpretation *of the sentence*. It is only the surrounding context that leads one to prefer an alternate interpretation, one that does involve the whole known civilized world. But what *precisely* should this interpretation be? For that we must consider some specifics of the context. Immediately after making the comment in question, Xunzi quotes an ode, wish reads: "The virtuous exemplary person, in decorum he does not err, in decorum he does not err, he rectifies the four countries." I do not dispute John Knoblock's interpretation that "the four countries" refers to "'the countries of the four regions,' that is, the whole world" (Knoblock 1990, p. 310 n136). But, what does it *mean* to "rectify" the world? And how does this relate to unifying the people? The above-quoted ode provides a clue: One rectifies the world by being flawless in decorum (*yi* 儀).

A bit earlier in the same section we find corroboration and clarification of this view that will help us sharpen our answer to our underlying question: What kind of global order is Xunzi imagining when he talks of unifying the people and rectifying the world? Xunzi writes:

[The person of genuine-humanity] would attend to what was appropriate for small and large, strong and weak [states], and thereby carefully preserve them. In maintaining ritualized distinctions, he would exhibit excellent form.... His persuasions surely would be cultured and refined, displaying the distinctive wisdom of an exemplary person" (*Xunzi* 10/48/14; 10.14).

Two important points can be gleaned from this passage. First, it corroborates the point made in the ode, namely, that proper decorum in diplomacy is key.

Second, it clarifies the world order that Xunzi is imagining, and thus guides our understanding of what rectification and unification must mean in this context. To repeat, Xunzi's person of genuine-humanity would "attend to what was appropriate for small and large, strong and weak [states], and thereby carefully preserve them." So, he is not imagining a political unification, but rather, orderly relations between independent states.

In the end, to "rectify the four countries" involves first a defensive strategy of maintaining the harmonious order within one's own humanely and intelligently governed state, and thereby maintaining security from those who desire to attack or take advantage. For they could never profit against a unified, prosperous, and orderly state, in which "the three armies are a unified force" (*Xunzi* 10/49/6; 10.14; and 10/48/12; 10.15). That was the main emphasis. But also, through decorum and diplomacy, engaged in from a position of both moral strength and strength of morale, an international order among individual states could be maintained as well, with attention to ritualized norms of propriety. This, then, is what the passage we started with means, namely, "The person of genuine-humanity (*ren ren* 仁人) uses his state not simply to hang onto what he has and no more, but to unify [all] people" (*Xunzi* 10/48/17; 10.14). He unifies them not under a single political order, but within a stable international system that is regulated by propriety between nations, as insisted upon by the humane leaders of well-ordered and secure states. One point that is not made explicit, but which I regard as a reasonable conjecture is this: The "rectification" of other states comes in part from the humane and authoritative ruler setting a moral and practical example. All the other rulers want what he (or she) is able to achieve. So, they are brought around, "rectified," by the persuasive practical example of the ruler as an exemplary person, as well as the power of his/her impressive good form.

The last thing I would note is that the view of international harmony that I have attributed to Xunzi here is similar to the view that Sungmoon Kim (2010) has persuasively attributed to Mencius. So, there are reasons to think that this view, or something like it, is one that modern Confucians should consider, even if one disagrees with my revised view of Xunzi's vision of world order.

OUR SITUATION AND OPTIONS

Let us review. Xunzi's point about the importance of "good people," and the need for a system of norms and roles designed to achieve this, is well taken. But the monolithic structure of "Xunzian global order" is too dangerous for humanity at its present state of cultural and spiritual evolution. We are too much like Xunzi describes, with our strong tendencies toward petty selfishness, to risk entrusting ourselves to a global regime—run by people like us—with no external checks or balances. Even if those devising the system were well-motivated and even if the system itself was well conceived (both of which are dubious assumptions), the risk that it may in time devolve into a stable tyranny militates strongly against the wisdom of pursuing that strategy at this time.

Mencius provides an alternative model, and it may be that Xunzi's view was actually not so different. In this model, there is an international federation of largely politically autonomous states, within a kind of moral hierarchy. Each state would concentrate on practicing benevolent government within its own sphere, which is consistent with the Confucian notion of graduated care, and each state would treat other nations as younger or elder brothers as appropriate to circumstances. Whether a given state had the role of younger or elder would at least in part be determined by a kind of moral competition. While this may sound somewhat naïvely idealistic, it may be useful to compare it with three alternatives:

1. Empire, like that sought by the PNAC group, and dreamt of by tyrants throughout history. Imperial projects are transparently immoral and *for this reason* eventually collapse. This is not an acceptable course.
2. The Westphalian conception of independent states, which is a contingent product of European history, and is roughly the status quo. This arrangement is precarious, as various predicaments around the world show. We are searching for some *improvement* from this condition.
3. Full-fledged globalization, wherein previously independent states are merged, for all practical purposes, into a global governing entity. I have argued that this option is too dangerous. I would add here that, in the eyes of those who value diversity as well as a reasonable measure of self-determination, it is also inherently unattractive.

In comparison, the option of Mencius's suzerain internationalism, in which individual states largely maintain political independence, seems relatively attractive. Further, since Mencius envisions a world that is in some sense unified, but not a monolith, it approximates our current institutional condition. And so, it is a vision that does not require radical changes in governing institutions. And yet it does call for moral improvement. This improvement would be a process that would be undergone in degrees, and there would always be a need to make adjustments to evolving conditions. It would involve a process in which states "muddle through" circumstances and leaders endeavor to act appropriately, and to live up to their potential as humans.

Yan Xuetong considers what would result if China actually lived up to its own Confucian standards. He writes:

> If China becomes a true kingship country—a super state grounded in high morals and ethics—it should bring about a world order more peaceful and secure than that today. True kingship may not be the perfect international system but, compared with the current hegemonic system, would be one imbued with greater cooperation and security. (Yan 2008, p. 159)

That is probably true. However, to cite just one problem, China cannot be a world leader and *at the same time* retain the excuse that if it allows free speech it might collapse into chaos. Where political stability is both fragile and of paramount importance, it may be morally acceptable to have stricter restrictions on free speech than would be most conducive to harmony in more stable circumstances. However, one cannot claim to be a world *leader* with a philosophy that is *worthy of emulation*, while simultaneously claiming to be *so fragile* as to be endangered by a few dissenting voices. Between these two options, (1) that it is ready for leadership, or (2) that it remains necessary to stifle dissent, China must choose. I hope it chooses morally responsible leadership.

Ultimately China's choices are up to China. As an American I am, and should be, chiefly concerned that my own country lives up to *its* highest ethical potential. This is because, among other reasons, modeling good behavior is ultimately the best way to have a positive effect on others, as has been stressed repeatedly.[30]

Some people actually argue that the U.S. *has been* doing mostly good around the world. We *have been* striving to be *ren*—just look at our "humanitarian" interventions! For example, former U.S. Secretary of State, James Baker asserts, "Generally speaking, when America is involved internationally we have been a force for peace and stability. Most countries know that we don't want their territory, we don't want their resources, and so forth."[31] But this claim does not survive scrutiny. Indeed, to believe this one must ignore most of the history of U.S. interventions during the past half-century or more, both overt and covert, in Latin America and around the world. From the overthrow of Iranian Prime Minister Mohammad Mosaddegh in 1953 to the invasion of Iraq[32] fifty years later, and so much unnecessary carnage in between—not to mention the continuation of current misadventures—the U.S. has acted shamefully in the pursuit of petty advantages and profits, often supporting "friendly" dictators while undermining democratically elected governments. *This* is what must change. But it cannot change without rigorous self-honesty, and careful monitoring of our leaders, remonstrating with them when appropriate, and insisting they lead with virtue.

NOTES

1. The following is one of the most chilling statements in *Rebuilding America's Defenses*: "[A]dvanced forms of biological warfare that can 'target' specific genotypes may transform biological warfare from the realm of terror to a politically useful tool" (Donnelly 2000, p. 60).

2. Regarding the expansion of executive branch powers, Chalmers Johnson explains, "After the attacks of September 11, 2001, cynical and shortsighted political leaders in the United States began to enlarge the powers of the president at the expense of the elected representatives of the people and the courts. The public went along, accepting the excuse that a little tyranny was necessary to protect the population" (Johnson 2006, p. 89).

3. It is characteristic of Confucians to believe that while petty governance would fail, altruistic governance would result in a harmonious situation, that is, one in which both the ruler and the people would benefit. See, *Xunzi* 9/36/10-11; 9.4, quoted on p. 145.

4. The importance of this distinction is discussed in Herman Daly's article "Globalization and its Discontents" (2001). Daly argues that, for a number of reasons not addressed in this chapter, internationalization is preferable to globalization.

5. Xunzi writes: "People are born with desires. If these desires are not fulfilled, [the object of desire] will surely be sought after. If this seeking has no measure or bounds, contention will be inevitable. If there is contention, then there will be chaos, and if there is chaos, there will be difficulty and impoverishment. The ancient kings detested this chaos. Thus, they fashioned ritual (*li*) and propriety (*yi*), and thereby made divisions that nurture people's desires and provide for their satisfaction" (*Xunzi* 19/90/3; 19.1a).

6. Xunzi writes: "According with the progression of the four seasons, and apportioning the myriad things, widely benefits the whole world for no other reason than this: it achieves proper divisions (*fen*) and is appropriate (*yi*). . . . A ruler is good at grouping. If the grouping and the guiding discourse (*dao*) are mutually coherent, then the myriad things all receive what is suitable to them. [For example] the six domestic animals all will get to grow to maturity, and members of all classes will live out their full life span" (*Xunzi* 9/39/12; 9.16a).

7. Cf. "If a sincere mind applies a sense of appropriateness then practical coherence will result" (*Xunzi* 3/11/5; 3.9; cf. 9.3 and 10.4).

8. In the *Analects*, being *ren* 仁, is equated with taking on a burden (任, also pronounced *ren*). See *Analects* 8.7, quoted on p. 128 n23, above.

9. Xunzi writes, "If one desires to achieve a harmonious and unified world, restraining [the large militaristic states] Qin and Chu, then no one can match a wise and insightful exemplary person (*junzi*)" (*Xunzi* 11/53/10; 11.7a).

10. See *Xunzi* 10/43/19-20; 10.5, quoted at length on page 135, where the sense of "persons of *ren*" (*ren ren*) is spelled out as "those who do their utmost in the service of others." Cf. *Xunzi* 10/48/17; 10.14, quoted on page 141.

11. Cf. Mencius says, "When force is used to make people submit, it is not that they submit in their hearts, it is that their strength [to resist] is inadequate. When

virtue (*de*) is used to make people submit, they are joyful in their hearts and submit sincerely" (*Mencius* 2A3).

12. Kant explains that all people need to do is: "create a good organisation for the state, a task which is well within their capability, and to arrange it in such a way that their self-seeking energies are opposed to one another, each thereby neutralising or eliminating the destructive effects of the rest. . . . As hard as it may sound, the problem of setting up a state can be solved even by a nation of devils (so long as they possess understanding)" (Reiss and Nisbet 1991, p. 112).

13. The clearest statement of this principle comes from the *Analects*, in which Confucius says. "Lead them with [legalistic] government, keep them in order with punishments, and the common people will avoid trouble but have no sense of shame. Lead them with virtue (*de*), keep them in order with ritual propriety (*li*), and they will not only develop a sense of shame, but will reform themselves" (*Analects* 2.3; cf. *Analects* 13.6 and 13.13).

14. Xunzi elaborates: "By training them, [sages] transformed people's emotional dispositions and guided them, causing everyone to move towards order, and to cohere with *dao*. Nowadays, people who are transformed by teachers and models, who accumulate culture and learning, and who are guided (*dao*) by ritual and propriety become exemplary persons (*junzi*). Those who indulge their dispositions and emotions, are self-satisfied and reckless, and go against ritual and propriety—they become inferior people" (*Xunzi* 23/113/12; 23.1b).

15. The lines immediately preceding this quotation read: "There are disorderly rulers, not disorderly states. There are orderly people, not orderly laws and norms (*fa*). The methods of the archer Yi are not lost, yet such an archer does not appear generation after generation. The laws and norms of the sage king Yu [of the Xia dynasty] still exist, yet the Xia could not continue its rule. Thus, laws and norms cannot stand alone. Categories cannot apply themselves."

16. Similarly, Confucius is reported to have said, "If the proper people are present, proper governance will be upheld. When these people are gone, proper governance comes to an end" (*Zhong Yong* 20).

17. Russell writes, "If raw materials are not to be used up too fast, there must not be free competition for their acquisition and use but an international authority to ration them. . . . This authority should deal out the world's food to the various nations. . . ." (Russell 1953, p. 111).

18. In scientific dictatorships of the future, explains Russell, "Diet, injections, and injunctions will combine, from a very early age, to produce the sort of character and the sort of beliefs that the authorities consider desirable, and any serious criticism of the powers that be will become psychologically impossible" (Russell 1953, p. 50).

19. Following his argument, Russell writes, "For these reasons, I do not believe that dictatorship is a lasting form of scientific society—unless (but this proviso is important) it can become world-wide" (Russell 1953, p. 55).

20. According to Supreme Court justice Elena Kagan, "Cass Sunstein is the preeminent legal scholar of our time—the most wide-ranging, the most prolific, the most cited, and the most influential" (Mangan 2008).

21. For my responses to Sunstein and Vermeule's article, see Hagen 2010a, 2011a.

22. Ford Explains, "The Chinese tradition has as its primary model for interstate relations a system in which the focus of national policy is, in effect, a struggle for primacy, and legitimate, stable order is possible only when one power reigns supreme—by direct bureaucratic control of the Sinic geographic core and by at least tributary relationships with all other participants in the world system. This monist model of global order is not merely a by-product of China's ancient history. Its central assumptions—about the need for political unity, the natural order of all politics as a pyramidal hierarchy, and the fundamental *legitimacy* of truly separate and independent state sovereignties—are reflected in many aspects of China's classical canon: in Confucian literature, Taoist works, and the manuals of war and statecraft known as the *bingjia*. Sinic monism, therefore, enjoys powerful roots in China's intellectual tradition that amplify its centrality as a prism through which all subsequent Chinese leaders have viewed their world and China's place in it" (Ford 2010, p. 4).

While I agree with Ford about the emphasis on unity in early Chinese thinking, Ford sometimes overstates it. There are examples that suggest that some value is placed on states maintaining peaceful co-existence with other states. For example, the *Laozi* has been plausibly interpreted as advice for rulers of small states (Graham 1989, p. 234, cf. Moeller 2004, pp. 6-8), to help them maintain their independence. In addition, Zhuangzi is not interested in governing at all, preferring to wander freely.

23. Similarly, one could argue that even a Xunzian strategy should not focus on globalizing structures, but instead on "cultivating the way," and doing right by others. One could, for example, cite the following passage: "King Tang and King Wu did not seize the world. They cultivated the way (*dao*) and put a sense-of-appropriateness (*yi*) into practice. They promoted the common benefit of the world, and eliminated the scourges of the world uniformly, and so the whole world turned to them" (*Xunzi* 18/84/6; 18.2).

24. Kim writes, "[Mencius] subscribed to the idea that international morality (and justice) is quite compatible with, and can only be sustained under, what I call the 'Confucian international moral hierarchy' among the states" (Kim 2010, p. 35, cf. p. 33).

25. Here I have used D. C. Lau's translation, as it appears in Kim 2010, p. 40. Bracketed clarifications are my own but are based on Kim's interpretation.

26. Mencius says, "*Tian* does not speak, it simply instructs through its motions and occurrences" (*Mencius*, 5A5; cf. *Mencius* 4A7, and 5A6).

27. Regarding the connection between benevolence and being genuinely human, Mencius says that to be *ren* 仁 (humane/benevolent) is to be *ren* 人 (human) (*Mencius* 7B16).

28. In one previously cited passage, which might be thought to imply the ideal of a politically unified world, Xunzi writes, "Exalting one [results in] order; exalting two [results in] chaos" (*Xunzi* 14/67/17; 14.7; cf. 9.3; quoted more fully on p. 134). It might be thought that this passage suggests, by extension, that there should be just one ruler of the whole world as well. But, if there is no problem with having multiple families, each with one father, then there should likewise be no problem having multiple states, each with one ruler. So, upon reflection, it is not at all clear that this passage implies that there should be a single ruler of a unified world order.

29. For example, John Knoblock translates the key phrase as "unite all peoples" (1990, p. 137).

30. See especially the list of quotations at the beginning of the "Introduction," pp. ix–x.

31. From an interview with James Baker in the dramatic-documentary-musical *American Ruling Class*.

32. See Bugliosi 2008 and deHaven-Smith 2010.

Conclusion

In contrast to the Western romanticization of war, the prominent early Chinese attitude toward warfare, as expressed in the central philosophical texts as well as military treatises, was that war was unfortunate, lamentable, tragic, and associated with loss even for the victor. War, from this perspective, should be avoided if possible and minimized if not. It certainly should not be glorified the way it has been in the West.

The early Confucian perspective on war can be best understood by taking into account rival philosophies of the same period. For example, the Mohists, being activists at least as much as philosophers, sought to discourage war by suggesting standards for the appropriate use of the military that were impossible to satisfy. Similarly, though Confucian standards were not as outlandish, the early Confucians put forward just war criteria that were exceedingly difficult to meet, and if they were met then military action would not amount to genuine war. In this way, they attempted to restrain the aggressive tendencies of ambitious rulers. In addition, like their Daoist rivals, early Confucians recognized that negative feedback loops would undermine efforts to use force to produce positive results. They also recognized the reliable benefits that flowed from benevolent governance and the wisdom of leveraging the virtuous cycles associated with doing good in morally unproblematic ways. This provided a basis for a strategy for rulers to increase their influence peacefully while having positive effects.

A number of scholars, however, have recently framed early Confucians as just war theorists with fairly permissive criteria. I argue that such permissive versions of the Confucian criteria have not been well established. And further, the *practical aim* of the early Confucians, who are seen as implicitly suggesting just war criteria, is to *prevent* war in their present circumstances by appealing to contrasts with idealized cases, not to provide low bars for future leaders to claim to have satisfied. Taking Confucian just war criteria seriously leads to the practical conclusion that offensive war, including interventions for humanitarian reasons (when that would involve the carnage of genuine war), is not justified in real or plausible circumstances. While some suggest that a relatively pacifistic reading of early Confucians interprets them as naïve, I've argued that this is not the case, especially when one considers the historical successes of non-violence and the failures of (ostensibly) well-meaning military interventions.

More specifically, I argue that neither Mencius nor Xunzi would support what we call "humanitarian interventions," at least not in plausible circumstances, and not when such interventions would result in genuine war and the attendant death and destruction. For the use of the military to be appropriate, in their view, the authority leading the intervention would have to be a person of *ren* 仁, or an agent of *tian* 天, which are exceptionally high standards. Indications that such a standard has been met include an overwhelming outpouring of support for the intervention on the part of the people of the intervened state. The intervention also must be universally supported by the leaders of neighboring states. These circumstances almost never occur, and if they did occur then successful military intervention would not require the carnage of genuine war.

The best way to deal with foreign despotism, from the early Confucian perspective, is by making one's own state all the more clearly the better one, increasing one's moral status and thus one's soft power. While this cannot generally bring a quick end to a crisis, over time problems can be ameliorated and conditions improved. It is a long-term strategic approach. While it may be hard to hold back from intervening when people are being harmed, Confucians understood what Daoists emphasized: violently forceful efforts are likely to backfire. And, there are significant limits on the good that violence can do. At the same time, there are always underemphasized opportunities to reduce suffering *somewhere* in ways that do not require morally problematic means.

Further, Confucians, especially Xunzi, recognized the reality of human selfishness and understood that most leaders were not immune from its influence. They knew that the rulers of their time often waged unjust wars, and that they would make excuses to try to frame their violence in a positive light. We know that this applies in our time too, a time in which disingenuous arguments for violence are frequently made. This knowledge bolsters the judgement that the Confucian skepticism regarding war, and their insistence on a high bar with which to judge arguments for the use of the military was not naïve. Indeed, it is imperative that scholars, officials, journalists, and other serious people look critically at arguments in support of violence and challenge these arguments if they are found to be faulty or seem to be motivated more by ideology, aggrandizement, or self-interest than by a prudently cautious aspiration to follow reliable ethical guidelines. Honest and critical focus on what is really going on, as distinct from how issues may be portrayed by interested parties, may help us avoid an unnecessary "clash of civilizations."

One way of putting an end to war, some people suggest, is by putting an end to independent sovereign nations states, replacing them with a single authority having a monopoly on military force. A plausible reading of Xunzi suggests that he would favor such benevolent globalism. That might be good if we could trust such unilateral power to exercise benevolence, and to do so indefinitely. But if such power were to turn oppressive, there would be nowhere to run. And there would be no moral competition, by virtue of which a better way could be demonstrated. Mencius provides a safer alternative, one in which a relatively harmonious system of largely independent states competes for moral status and for the respectability and soft power that goes with it. Upon reconsideration, there is also a plausible reading of Xunzi under which he largely shares this vision. It is a vision worth considering.

Bibliography

Ackerman, Peter and Jack DuVall. 2000. *A Force More Powerful: A Century of Nonviolent Conflict*. New York: St. Martin's Press.
Ames, Roger T. 1993. *Sun-Tzu The Art of Warfare*. New York: Ballantine Books.
Ames, Roger T. and David L. Hall. 2003. *Daodejing: "Making This Life Significant."* New York: Ballantine Books.
Ames, Roger T. and Henry Rosemont, Jr. 1998. *The Analects of Confucius: A Philosophical Translation*. New York: Ballantine Books.
Anscombe, G. E. M. 1958. "Modern Moral Philosophy." *Philosophy* 33.124.
Arendt, Hannah. 1970. *On Violence*. New York: Harcourt, Brace and World.
Bai, Tongdong. 2012. *China: The Political Philosophy of the Middle Kingdom*. New York: Zed Books.
Ball, Howard. 2007. *Bush, the Detainees, and the Constitution: The Battle over Presidential Power in the War on Terror*. Lawrence, Kansas: University Press of Kansas.
Bawden, Charles R. 2020. "Kublai Khan." *Encyclopædia Britannica,* published January 1, 2020, accessed December 29, 2020. https://www.britannica.com/biography/Kublai-Khan
Bell, Daniel A. 2008. "Just War and Confucianism: Implications for the Contemporary World." In *Confucian Political Ethics*, edited by Daniel Bell, 226-256. Princeton, NJ: Princeton University Press.
Berling, Judith A. 2004. "Confucianism and Peace Building." In *Religion and Peacebuilding,* edited by Harold Coward and Gordon Smith, 93-110. Albany: SUNY University Press.
Bugliosi, Vincent. 2008. *The Prosecution of George W. Bush for Murder*. New York: Vanguard Press.
Byrne, Rebecca. 1974. *Harmony and Violence in Classical China: A Study of the Battles of the Tso-Chuan*. University of Chicago Ph.D. dissertation.
Cady, Duane. 2010. *From Warism to Pacifism: A Moral Continuum* (2nd edition). Philadelphia: Temple University Press.

Calhoun, Laurie. 2002. "How Violence Breeds Violence: Some Utilitarian Considerations." *Politics* 22.2: 95-108.
Chan, Joseph. 2008. "Confucian Attitudes toward Ethical Pluralism." In *Confucian Political Ethics*, edited by Daniel A. Bell, 113-138. Princeton, NJ: Princeton University Press.
Chan, Joseph. 2014. *Confucian Perfectionism: A Political Philosophy for Modern Times*. Princeton, NJ: Princeton University Press.
Chen, Frederick Tse-shyang. 1999. "The Confucian View of World Order." In *Religion and International Law*, edited by Mark W. Janis and Carolyn Evans, 27-49. Cambridge, MA: Kluwer Law International.
Chenoweth, Erica and Maria Stephan. 2011. *Why Civil Resistance Works: The Strategic Logic of Nonviolent Conflict*. New York: Columbia University Press.
Ching, Julia. 2004. "Confucianism and Weapons of Mass Destruction." In *Ethics and Weapons of Mass Destruction: Religious and Secular Perspectives*, edited by Sohail H. Hashmi and Steven P. Lee, 246-269. New York: Cambridge University Press.
Clausewitz, Carl von. 1968. *On War*. Edited by Anatol Rapoport. New York: Penguin Books.
Coady, David. 2006. *Conspiracy Theories: The Philosophical Debate*. Burlington, VT: Ashgate.
Coady, David. (ed.) 2007. *Episteme: A Journal of Social Epistemology* 4.2 (Special Issue: Conspiracy Theories).
Daly, Herman E. 2001. "Globalization and its Discontents." *Philosophy and Public Policy Quarterly* 21.2/3 (Spring/Summer 2001): 17-21.
DeHaven-Smith, Lance. 2010. "State Crimes Against Democracy in the War on Terror: Applying the Nuremberg Principles to the Bush-Cheney Administration." *Contemporary Politics* 16.4 (December): 403-420.
Dentith, M R. X., ed. 2018. *Taking Conspiracy Theories Seriously*. Lanham, MD: Rowman & Littlefield.
Donnelly, Thomas. 2000. *Rebuilding America's Defenses: Strategy, Forces and Resources*. The Project for the New American Century (PNAC).
Doward, Jamie, Gaby Hinsliff, and Mark Townsend. 2009. "Confidential Memo Reveals US Plan to Provoke an Invasion of Iraq." *The Observer*, June 20, 2009. https://www.theguardian.com/politics/2009/jun/21/iraq-inquiry-tony-blair-bush
Fairbank, John K. 1974. "Introduction: Varieties of the Chinese Military Experience." In *Chinese Ways in Warfare*, edited by Frank Kierman and John K. Fairbank, 1-26. Cambridge, MA: Harvard University Press.
Ferguson, John. 1978. *War and Peace in the World's Religions*. New York: Oxford University Press.
Fingarette, Herbert. 1972. *Confucius: The Secular as Sacred*. New York: Harper Torchbooks.
Finn, Peter and Sari Horwitz. 2012. "Holder: U.S. Can Lawfully Target American Citizens." *Washington Post*, March 5, 2012. https://www.washingtonpost.com/world/national-security/holder-us-can-lawfully-target-american-citizens/2012/03/05/gIQANknFtR_story.html

Fisher, Louis. 2008. *The Constitution and 9/11: Recurring Threats to America's Freedoms*. Lawrence, KS: University Press of Kansas.

Ford, Christopher. 2010. *The Mind of Empire: China's History and Modern Foreign Relations*. Lexington: University Press of Kentucky.

Fraser, Chris. 2016. *The Philosophy of the Mozi: The First Consequentialists*. New York: Columbia University Press.

Goldin, Paul Rakita. 2011. *Confucianism*. Durham, UK: Acumen.

Graham, A. C. 1989. *Disputers of the Tao: Philosophical Argument in Ancient China*. La Salle, IL: Open Court.

Hagen, Kurtis. 1996. "A Chinese Critique on Western Ways of Warfare." *Asian Philosophy* 6.3 (November): 207-217.

Hagen, Kurtis. 2007. *The Philosophy of Xunzi: A Reconstruction*. La Salle, IL: Open Court.

Hagen, Kurtis. 2010a. "Is Infiltration of 'Extremist Groups' Justified?" *International Journal of Applied Philosophy* 24.2: 153-168.

Hagen, Kurtis. 2010b. "The Propriety of Confucius: A Sense-of-Ritual." *Asian Philosophy* 20.1 (March): 1-25.

Hagen, Kurtis. 2011a. "Conspiracy Theories and Stylized Facts." *Journal for Peace and Justice Studies* 21.2: 3-22.

Hagen, Kurtis. 2011b. "Xunzi and the Prudence of *Dao*: Desire as the Motive to Become Good." *Dao: A Journal of Comparative Philosophy* 10.1 (March): 53-70.

Hagen, Kurtis. 2016a. "Project for a New Confucian Century." In *A Future Without Borders: Theories and Practices of Cosmopolitan Peacebuilding*, edited by Eddy Souffrant, 168-189. New York: Brill/Rodopi.

Hagen, Kurtis. 2016b. "Would Early Confucians Really Support Humanitarian Interventions?" *Philosophy East and West* 66.3 (July): 818-841.

Hagen, Kurtis. 2018a. "Conspiracy Theories and Monological Belief Systems." *Argumenta* (Journal of the Italian Society for Analytic Philosophy), special issue on the ethics and epistemology of conspiracy theories.

Hagen, Kurtis. 2018b. "Conspiracy Theories and the Paranoid Style: Do Conspiracy Theories Posit Implausibly Vast and Evil Conspiracies?" *Social Epistemology* 32.1: 24-40.

Hagen, Kurtis. 2018c. "Conspiracy Theorists and Social Scientists." In *Taking Conspiracy Theories Seriously*, edited by M. R. X. Dentith, 125-140. London: Rowman and Littlefield.

Hagen, Kurtis. 2019. "From Patterning to Governing: A Constructivist Interpretation of the *Xunzi*." In *The Bloomsbury Research Handbook of Early Chinese Ethics and Political Philosophy*, edited by Alexus McLeod, 45-65. New York: Bloomsbury.

Hagen, Kurtis. 2020a. "Should Academic Debunk Conspiracy Theories?" *Social Epistemology* 34.5: 423-439.

Hagen, Kurtis. 2020b. "Is Conspiracy Theorizing Really Epistemically Problematic?" *Episteme*, 1-23. doi:10.1017/epi.2020.19.

Hagen, Kurtis (forthcoming). *Conspiracy Theories and the Academy: A Philosophical Critique*. Ann Arbor, MI: University of Michigan Press.

Hagen, Kurtis, and Steve Coutinho. 2018. *Philosophers of the Warring States: A Sourcebook in Chinese Philosophy*. Peterborough, Ontario: Broadview Press.

Hall, David L., and Roger T. Ames. 1987. *Thinking Through Confucius*. Albany, NY: SUNY Press.

Harris, Eirik Lang. 2019. "Xunzi on the Role of the Military in a Well-Ordered State." *Journal of Military Ethics* 18.1: 48–64.

Hart, H. B. Liddell. 1963. "Forward." In *Sun Tzu, The Art of War*, translated by Samuel B. Griffith. New York: Oxford University Press.

Hayward, Tim. 2019. "Three Duties of Epistemic Diligence." *Journal of Social Philosophy* 50.4: 536–561.

Holder, Eric H. 2013. (Letter to Patrick J. Leahy, May 22). http://www.justice.gov/ag/AG-letter-5-22-13.pdf.

Holmes, Robert L. 2013. *The Ethics of Nonviolence: Essays by Robert L. Holmes*. Edited by P. Cicovacki. New York: Bloomsbury.

Huang, Pumin. 1996. "Xianqin Zhuzi Junshi Sixiang Yitong Chutan" [A Preliminary Comparative Study of the Military Thought of the Pre-Qin Masters], *Lishi Yanjiu* [Historical Research], no. 5.

Hutton, Eric L. 2014. *Xunzi: The Complete Text*. Princeton, NJ: Princeton University Press.

Ivanhoe, Philip J. 1991. "A Happy Symmetry: Xunzi's Ethical Thought." *Journal of the American Academy of Religion* 59.2: 309-322.

Ivanhoe, Philip J. 2000. "Human Nature and Moral Understanding in the *Xunzi*." In *Virtue, Nature, and Moral Agency in the Xunzi*, edited by T. C. Kline III and Philip J. Ivanhoe, 237–49. Indianapolis: Hackett.

Ivanhoe, Philip J. 2004. "'Heavens Mandate' and the Concept of War in Early Confucianism." In *Ethics and Weapons of Mass Destruction: Religious and Secular Perspectives*, edited by Sohail H. Hashmi and Steven P. Lee, 270-276. New York: Cambridge University Press.

Ivanhoe, Philip J. 2011. "Introduction." In *Master Sun's Art of War*, translated with introduction, by P. J. Ivanhoe, xiii-xxx. Indianapolis: Hackett.

Ivanhoe, Philip J. and Bryan W. Van Norden, eds. 2001. *Readings in Classical Chinese Philosophy*, 2nd edition. Indianapolis: Hackett.

Jackson, Richard. 2020. "A Pacifist Critique of Just War Theory." In *Comparative Just War Theory: An Introduction to International Perspectives*, edited by Luis Cordeiro-Rodrigues and Danny Singh, 45-59. Lanham, MD: Rowman & Littlefield.

Johnson, Chalmers. 2006. *Nemesis: The Last Days of the American Republic*. New York: Metropolitan Books.

Juergensmeyer, Mark. 2005. *Gandhi's Way: A Handbook of Conflict resolution* (revised edition). Berkeley: University of California Press.

Kim, Sungmoon. 2010. "Mencius on International Relations and the Morality of War: From the Perspective of Confucian *Moralpolitik*." *History of Political Thought* 31.1 (spring): 33-56.

Kim, Sungmoon. 2016. "Achieving the Way: Confucian Virtue Politics and the Problem of Dirty Hands." *Philosophy East & West* 66.1 (January): 152-176.

Kim, Sungmoon. 2017. "Confucian Humanitarian Intervention? Toward Democratic Theory." *The Review of Politics* 79: 187-213.

Klein, Naomi. 2007. *The Shock Doctrine: The Rise of Disaster Capitalism*. New York: Picador.

Knoblock, John. 1988, 1990, 1994. *Xunzi: A Translation and Study of the Complete Works* (three volumes). Stanford, CA: Stanford University Press.

Kupperman, Joel. 2007. "Mencius," in *Classic Asian Philosophy: A Guide to the Essential Texts*, 2nd edition. New York: Oxford University Press.

Lang, Andrew. 1921. *The Iliad of Homer*. London: Macmillan and Co.

Lau. D.C. 1970. *Mencius*. New York: Penguin Books.

Lau. D.C. 1979. *Confucius: The Analects*. New York: Penguin Books.

Lau, D. C., and F. C. Chen, eds. 1996. *A Concordance to the Xunzi*. Institute of Chinese Studies (ICS) Ancient Chinese Texts Concordance Series. Hong Kong: The Commercial Press.

Lau, D.C. and Roger T. Ames. 1996. *Sun Pin: The Art of Warfare*. New York: Ballantine Books.

Lepard, Brian D. 2002. *Rethinking Humanitarian Intervention: A Fresh Legal Approach Based on Fundamental Ethical Principles in International Law and World Religions*. University Park, PA: Pennsylvania State University Press.

Levin, Carl. 2011. "New Report: Bush Officials Tried to Shift Blame for Detainee Abuse to Low-Ranking Soldiers." *Huffington Post*, updated May 25, 2011. https://www.huffpost.com/entry/new-report-bush-officials_b_189823

Lewis, Charles, and Mark Reading-Smith. 2008. "Iraq: The War Card: False Pretenses." *Center for Public Integrity*, January 23, 2008. https://publicintegrity.org/politics/false-pretenses/

Liivoja, Rain. 2013. "Law and Honor: Normative Pluralism in the Regulation of Military Conduct." In *Normative Pluralism and International Law: Exploring Global Governance*, edited by Jan Klabbers, and Touko Piiparinen, 143-165. New York: Cambridge University Press.

Litvinenko, Alexander, and Yuri Felshtinski. 2007. *Blowing Up Russia: The Secret Plot to Bring Back KGB Terror*. New York: Encounter Books.

Lo, Ping-cheung. 2015a. "Varieties of Statecraft and Warfare Ethics in Early China: An Overview." In Lo and Twiss, 2015: 3-24.

Lo, Ping-cheung 2015b. "The *Art of War* Corpus and Chinese Just War Ethics Past and Present." In Lo and Twiss, 2015: 29-65.

Lo, Ping-cheung 2015c. "Warfare Ethics in Sunzi's *Art of War*? Historical Controversies and Contemporary Perspectives." In Lo and Twiss, 2015: 66-89.

Lo, Ping-cheung 2015d. "Legalism and Offensive Realism in the Chinse Court Debate on Defending National Security." In Lo and Twiss, 2015: 249-280.

Lo, Ping-cheung, and Sumner B. Twiss (eds.) 2015. *Chinese Just War Ethics: Origin, Development, and Dissent*. New York: Routledge.

Loy, Hui-chieh. 2015. "Mohist Arguments on War." In Lo and Twiss, 2015: 226-248.

MacAskill, Ewen. 2005. "Blix Insists There Was No Firm Weapons Evidence." *The Guardian*, April 28, 2005. https://www.theguardian.com/politics/2005/apr/28/iraq.iraq

Machiavelli, Niccolò. 1965. *The Art of War*. Translated by Ellis Farneworth. New York: Da Capo Press.

Mangan, Katherine. 2008. "Cass Sunstein to Leave U. of Chicago for Harvard," *The Chronicle of Higher Education*, February 19, 2008.

Mathies, David Kratz. 2011. "Should I Help the Empire with My Hand? Confucian Resources for a Paradigm of Just Peacemaking." *Journal of Religion, Conflict, and Peace* 4.2.

Mearsheimer, John J. 2014. *The Tragedy of Great Power Politics*. New York: Norton.

Mo, Di [Mozi]. 2010. *The Mozi: A Complete Translation*. Translated by Ian Johnston. New York: Columbia University Press.

Moeller, Hans-Georg. 2004. *Daoism Explained: From the Dream of the Butterfly to the Fishnet Allegory*. La Salle, IL: Open Court.

Moeller, Hans-Georg. 2006. *The Philosophy of the Daodejing*. New York: Columbia University Press.

Needham, Joseph. 1969. *The Grand Titration; Science and Society in East and West*. Toronto: University of Toronto Press.

Ni Lexiong. 2008. "The Implications of Ancient Chinese Military Culture for World Peace." In *Confucian Political Ethics*, edited by Daniel Bell, 201-225. Princeton, NJ: Princeton University Press.

Noddings, Nel. 2006. "The Psychology of War." *Critical Lessons: What Our Schools Should Teach*. New York: Cambridge University Press.

Orwell, George. 2008. *Nineteen Eighty-Four*. Boston: Houghton Mifflin Harcourt.

Qin, Cao, 2020. "The Classical Confucian Ideas of *Jus ad Bellum*." In *Comparative Just War Theory: An Introduction to International Perspectives*, edited by Luis Cordeiro-Rodrigues and Danny Singh, 157-171. Lanham, MD: Rowman & Littlefield.

Reiss, Hans (ed.) and H. B. Nisbet (trs.). 1991. *Kant: Political Writings*, 2nd edition. New York: Cambridge University Press.

Rowse, Arthur E. 1992. "How to Build Support for War." *Columbia Journalism Review* (September/October). http://web.archive.org/web/20070824210500/http://backissues.cjrarchives.org/year/92/5/war.asp

Ruskin, John. 1964. "War." In *Man and Warfare: Thematic Readings for Composition*, edited by W.F. Irmscher, 35-61. Boston: Little, Brown and Company.

Russell, Bertrand. 1953. *The Impact of Science on Society*. New York: Simon and Schuster.

Scott, Peter Dale. 2007. *The Road to 9/11: Wealth, Empire, and the Future of America*. Berkeley: University of California Press.

Shue, Henry. 2009. "Making Exceptions." *Journal of Applied Philosophy* 26.3: 307–22.

Simons, Geoff. 1996. *The Scourging of Iraq: Sanctions, Law, and Natural Justice*. New York: St. Martin's Press.

Slingerland, Edward. 2003. *Confucius: Analects*. Indianapolis, IN: Hackett.

Stalnaker, Aaron. 2015. "Xunzi's Moral Analysis of War and Some of its Contemporary Implications." In Lo and Twiss, 2015: 135-152.

Stroble, James A. 1998. "Justification of War in Ancient China." *Asian Philosophy* 8.3.

Sunstein, Cass. and Adrian Vermeule. 2009. "Conspiracy Theories: Causes and Cures." *The Journal of Political Philosophy* 17.2: 202-227.

Twiss, Sumner B. and Jonathan Chan. 2015a. "The Classical Confucian Position on the Legitimate Use of Military Force." In Lo and Twiss, 2015: 93-116.

Twiss, Sumner B. and Jonathan Chan. 2015b. "Classical Confucianism, Punitive Expeditions, and Humanitarian Intervention." In Lo and Twiss, 2015: 117-134.

Valentine, Douglas. 1990. *The Phoenix Program*. Lincoln, NE: iUniverse.

Van Els, Paul. 2013. "How to End Wars with Words: Three Argumentative Strategies by Mozi and His Followers." In *The Mozi as an Evolving Text: Different Voices in Early Chinese Thought*, edited by Carine Defoort and Nicolas Standaert, 69–94. Leiden: Brill.

Van Norden, Bryan W. 1997. "Mencius on Courage." *Midwest Studies in Philosophy* 21: 237-256.

Van Norden, Bryan W. (trans.) 2008. *Mengzi: With Selections from Traditional Commentaries*. Indianapolis, IN: Hackett.

Waley, Arthur. 1939. *Three Ways of Thought in Ancient China*. London: George Allen & Unwin.

Walter, J. H. 1954. *The Arden Edition of the Works of William Shakespeare: King Henry V*. New York: Methuen.

Wang Liqi (王利器). ed. 1992. *Critical Edition of the Discourses on Salt and Iron with Annotations: Finalized Edition*. 《鹽鐵論校注》(定本). Beijing: Zhonghua Book Company.

Watson, Burton. 1964. *Chuang Tzu: Basic Writings*. New York: Columbia University Press.

Watts, Alan. 1966. *The Book on the Taboo Against Knowing Who You Are*. New York: Vintage Books.

Whitehouse, Sheldon. 2009. *Congressional Record* (Senate) 155.85.42 (Torture), June 9, 2009: S6359-S6361. https://www.congress.gov/congressional-record/2009/06/09/senate-section/article/S6359-1

Wilkerson, Lawrence. 2009. "The Truth About Richard Bruce Cheney." *The Washington Note*, May 13, 2009. https://washingtonnote.com/the_truth_about/

Windrem, Robert and Victor Limjoco. 2008. "9/11 Commission Controversy." *Deep Background, MSNBC.com*, January 30, 2008.

Wood, Neal. 1965. "Introduction." In *The Art of War / Niccolò Machiavelli*. Translated by Ellis Farneworth. New York: Da Capo Press.

Xu, Fuguan 徐復觀. 1979. *Intellectual History of Han Dynasties*《兩漢思想史》(Vol. 3). Taipei: Xuesheng Press.

Yan Xuetong. 2008. "Xun Zi's Thoughts on International Politics and Their Implications." *Chinese Journal of International Politics* 2: 135-165.

Yan Xuetong. 2011. *Ancient Chinese Thought, Modern Chinese Power*. Princeton, NJ: Princeton University Press.

Yang Qianru. 2011. "An Examination of the Research Theory of Pre-Qin Interstate Political Philosophy." In *Ancient Chinese Thought, Modern Chinese Power*, edited by Daniel Bell and Sun Zhe, 147-160. Princeton, NJ: Princeton University Press.

Yao, Xinzhong. 2004. "Conflict, Peace and Ethical Solutions: A Confucian Perspective on War." *Sungkyun Journal of East Asian Studies* 4.2: 89-111.

Yu, Yi-Ming. 2016. "Military Ethics of Xunzi: Confucianism Confronts War." *Comparative Strategy* 35.4: 260-273.

Yuan, Baoxin 袁保新. 1992. *The Three Distinctions in Mencius: Historical Reflection and Modern Interpretation* 《孟子三辨之學的歷史省察與現代詮釋》. Taipei: Wenjin Press.

Zhang, Ellen. 2015a. "'Weapons are Nothing but Ominous Instruments': The *Daodejing*'s View on War and Peace." In Lo and Twiss, 2015: 181-208.

Zhang, Ellen. 2015b. "*Zheng* (征) as *Zheng* (正)? A Daoist Challenge to Punitive Expeditions." In Lo and Twiss, 2015: 209-225.

Ziporyn, Brook. 2009. *Zhuangzi: The Essential Writings with Selections from Traditional Commentaries*. Indianapolis: Hackett.

Index

activism, x–xi, 33–35, 39, 41, 43, 57, 64, 74, 109, 149
aggression, xii, 21, 34–36, 57, 81, 88n1, 94, 102–3, 139
America, xiii, 17, 28–29, 62, 69, 85, 114–115, 118–21, 123–25, 131–33, 144
Ames, Roger T., viii, 8, 20, 22, 24, 26, 31n17
Analects (of Confucius), viii, ix, x, xii–xiii, 1–2, 4–8, 9n2, 19, 30n4, 30n9, 31n16, 42, 44, 48–50, 53n17, 69, 82, 88n6, 104, 109, 121, 123, 128n21, 128nn23–24, 145n8, 146n13
annexation, 31n14, 64–65, 105
Anscombe, G.E.M., 109
Arendt, Hannah, 37

Bai, Tongdong, 56–57, 67–70, 76n6, 77n19, 85
Bell, Daniel, 11, 45, 51–52, 56–57, 59–63, 65–69, 73, 76n7, 104–5, 112n14
Berling, Judith A., 46
Blair, Tony, 126n11
Blix, Hans, 126n11
Bush, George H. W., 66
Bush, George W., 62, 66, 115, 118–19, 126nn10–11, 129n28

Byrne, Rebecca, 24

Calhoun, Laurie, 37
Chan, Jonathan, xii, 55–57, 73, 76n7, 88n1, 91–94, 97–106, 108, 111nn1–4, 112n10, 112nn15–17, 134
Chan, Joseph, 6, 44
Chen, Frederick Tse-shyang, 56
Chenoweth, Erica, 53n16
Chinese Terms
  dao, the way, x, 7–9, 22, 25, 27, 36, 46, 50, 53n17, 54n19, 82, 84–86, 97, 123, 128n23, 135, 137, 145n6, 146n14, 147n23
  de, influential virtue, ix, 4, 19, 28, 38, 42, 44, 72, 83–85, 88n7, 129n26, 135, 146nn11,13
  fa, (role) model, xivn1,9, 42, 44, 46, 49, 115, 135, 144, 146n14
  jian'ai, impartial care, 1, 34
  junzi, exemplary person, x, 4–6, 8, 30n4, 44, 50, 88n7, 90n14, 116, 134, 141–42, 145n9, 146n14
  li, ritual propriety, ix–x, 1–8, 25, 27–28, 37, 44–48, 53n17, 84, 86, 89nn8–9, 101–2, 114–16, 133–36, 145n5, 146nn13–14
  quan, discretion, 45, 47–48

161

ren, benevolence/humanity, xiii, 2–5, 9n5, 19, 25, 44–45, 47, 50–52, 53n13, 57, 61, 65, 67, 70, 82–85, 89n12, 108, 125n1, 128n23, 134, 138, 140, 145n8, 147n27
ren ren, person of humanity, 5, 25, 50, 82–83, 90n12, 108, 134–35, 141–42, 150
shu, empathetic consideration, 5, 44
tian, the heavens/"Heaven," 7, 8, 27, 52n1, 64, 74, 139–40, 147n26, 150
wei, awe-inspiring majesty 84, 89n8, 90n13
wuwei, non-action, 31n11, 41–42
xing e, natural disposition are detestable, xii, 113–15, 128n22
yi, appropriateness, 2–4, 6, 9n5, 25–27, 44, 48, 50, 52n1, 53n13, 58, 81–82, 84–85, 87, 89n9, 95, 97, 108, 112n17, 116, 122, 132–33, 145nn5–6
zhong, doing one's utmost, 5, 44
Ching, Julia, 105–7, 111n6
chivalry. *See* gallantry
clash of civilizations, xii, 110, 113, 117, 122–23, 151
Clausewitz, Carl von, 13–14, 16, 24
clean hands critique, 17, 33, 46–49, 52
coercion, 31n17, 32n22, 38, 42–43 53n10, 87, 88n2, 131, 134. *See also* non-coercion
Confucian idealism, xi, 36–37, 39–40, 43, 134
Confucian solution, x, xii–xiii, 11, 28–29, 36–37, 42, 51–52, 85. *See also* governance
Confucius, xiii, 1–2, 4–8, 19, 38, 49–51, 56, 58–59, 65, 69, 75n2, 79–80, 121–23, 146n16. *See also Analects* (of Confucius)
consequences of war. *See* war, consequences of
conservative side of Confucianism, 9, 44, 86

conspiracy theories, xiii, xivn3, 119, 128n25. *See also* September 11, and Russian bombings
courage, 12–14, 16, 19–20, 22, 30nn4–5, 78n22, 94–95, 123
Coutinho, Steve, 52n1
cowardice, 16, 22

Da Xue (the *Great Learning*), 124, 129n26
Daly, Herman E., 145n4
Daoism, xi, 1, 11, 20–23, 31n16, 32n18, 33, 40–43, 47, 53n11, 149–50
desires, xii, 3, 5–6, 23, 41, 61, 69, 75, 114–16, 122, 125n3, 145n5
duty to fight. *See* humanitarian intervention, responsibility to protect

evil, 70, 80, 111n2;
versus good, x, 18, 68, 137;
necessary, 31n17, 81. *See also* tyrants
exceptionalism, 114–15, 118, 121, 144
expedient measures 17, 47. *See also* clean hands critique

Fairbank, John K., 56
fame, 12, 14–15, 20–21, 25, 53n15, 89n7, 116. *See also* honor, and reputation
feedback loops, 47, 93, 109, 149
Felshtinski, Yuri, 127n7
Ferguson, John, 75n2
flexibility regarding norms, xi, 43, 45–48, 128n24
Ford, Christopher, 138, 147n22
fortuna, 13
Fraser, Chris, 34–36

gallantry, 13–14, 16–19
Gandhi, 44, 54n20
governance, proper/benevolent, ix, xi, xiii, 3, 5, 36, 38, 44, 50, 64–65, 79–80, 93, 101–2, 107–9, 139–40, 145n3, 146n16, 149

Gulf Wars, 17, 28, 30n3, 62–63, 66, 72n12, 118–21, 126n11, 127n17, 129n27

Hall, David L., 8, 31n17
Han Dynasty 38
Han Feizi, 2
Harris, Eirik Lang, 36–37
Hart, H. B. Liddell, 24
Hayward, Tim, 90n15
hegemon / lord-protector, 19, 49, 66, 103–104, 112n13
heroism, 11–12, 14, 18, 20, 29n2, 30n8, 53n15
Holder, Eric H., 69, 77n17
Holmes, Robert L., 77n18
honor, 12–14, 17–18, 20, 25–26, 29n2, 48, 95, 97, 114, 116. *See also* fame
humanitarian assistance, non-military, 85, 87, 149
humanitarian intervention, xi, xii, 33, 40, 42, 45, 51–52, 55–57, 59–72, 79, 82–86, 90nn15–16, 91, 100, 102, 108–10, 150; responsibility to protect, 18, 76n7, 91, 93, 103–4, 108, 111n3; versus humane intervention, 99–100
Hutton, Eric L., 83, 88n4, 89nn10–11
Huxlian Brave New World, 134

ideal / non-ideal circumstances, 59–61, 70, 73–74, 84
idealized tales, xi, 38, 57, 59, 64, 72, 90n16, 93, 97–98, 108, 149
imperial project, 116, 120, 123–124, 131–133, 143
intellectuals, responsibility of, xiii, 109–110, 113, 116–17, 120, 122, 150–51
invincibility, 36, 108
Iraq war, *See* Gulf wars
Ivanhoe, Philip J., 20–21, 32n21, 105–6, 136–37

Jackson, Richard, 28, 32n23, 37, 52n1, 54n21, 90n16, 107, 111n2

Johnson, Chalmers, 145n2
just war criteria
  international support, 62–63, 66–67, 71–72, 105, 112nn14,16, 150
  as impossible to satisfy, 33–35, 52n4,62, 72, 86, 110–115, 149
  just cause, 30n7, 39, 92, 111n1
  lack of resistance, 55, 57, 60–61, 65, 72, 83–84, 97–98, 105, 110
  last resort, 69, 92
  likelihood of success, 60, 92, 105
  minimal bloodshed, 65, 101, 106
  moral authority of intervening ruler, 7, 25, 56, 58, 63–64, 72, 87, 100–104
  popular support, 25, 57, 59–60, 62, 68, 71–72, 89n11, 104, 110, 150
  proportionality, 92, 105
  right intention, 92, 98, 119
  serious misrule as justification, xii, 57, 91–93, 98, 100
justice, 32n23, 134;
  in Confucianism: 58, 67, 80–82, 88n1, 93, 100–103, 106, 108, 147n24;
  in Daoism 12, 41, 53n13
  weak versions of, 57, 67–69, 104

Kagan, Elena, 146n20
Kant, Immanuel, 134–135, 146n12
killing: in Confucianism, 7, 44, 48, 50–51, 54nn19,20, 67, 106; in Daoism, 23; in war, 19, 30n3, 32n17, 47, 60, 83
Kim, Sungmoon, 46–47, 70–72, 77–78nn20–21, 92, 99–100, 111n9, 112n14, 125, 138–39, 142, 147n24
Klein, Naomi, 126n12, 127n14
Knoblock, John, viii, 88n4, 89n11, 141, 148n29
Kublai Khan, 134
Kupperman, Joel, 140

Laozi, 1, 7, 12, 19–23, 30nn7–8, 31n11, 32n19, 40–43, 53nn13,15, 147n22

Index

Lau, D.C., viii, 20, 22, 30n4, 46, 96, 99
law, 6, 18, 77n17, 89n7, 115, 135, 146nn13,15
leadership, moral, ix–x, xiii, 2–4, 37–38, 42, 89n9, 115, 135, 144
legalism, 2, 38–39, 53nn9–10
Levin, Carl, 127n16
Liang Qichao, 75n2
Litvinenko, Alexander, 127n7, 127n9
Lo, Ping-cheung, 20–21, 31n12, 37–40, 52–53nn8–11, 56, 108–9, 111n3
long-term strategy, xiii, 11, 36–38, 42–43, 47, 57, 70
lord-protector. *See* hegemon / lord-protector
Loy, Hui-chieh, 21, 34–35, 52n3–4

Machiavelli, Niccolò, 12–14
Manchurian/Mukden Incident, 117, 125n6
Martin Luther King, 44
Mathies, David Kratz, 48–49, 51n18, 75n2
means-ends tension, xi, 27, 44–48, 51, 56, 80, 137
Mearsheimer, John J., 40
Mencian international harmony, xiii, 125, 131, 133, 138–40
Mencius, viii, x–xiv, 1–8, 9n5, 19–20, 25–26, 31n15, 36–38, 45–52, 53nn14,17, 54nn20–21, 55–65, 67–75, 75nn1–2, 76nn7–9,14–15, 77nn16–17,19, 78n22, 79–80, 82, 86, 88n1, 90n16, 91–109, 111nn1, 4–5,9, 112nn16–17, 114, 133, 138–40, 142–43, 145n11, 147n24, 26–27, 150–51
military campaigns, semi-mythic stories of: King Tang, 59, 64, 76n9, 85, 87, 88n5, 92, 96–97, 147n23; King Wen, 64, 88n5, 94–95; King Wu, 61, 64, 67, 76n9, 88n5, 94–95, 111n6, 147n23. *See also* idealized tales
Moeller, Hans-Georg, 11–12, 19, 23, 30nn7–8, 32n18, 40–41, 53n15

Mohism, x–xi, 1, 8, 11, 19, 21–22, 29, 33–36, 41, 52nn1,3–4
moral suasion, 38, 108, 134, 141
moralpolitik, 138–140
Mozi, 1, 8, 12, 21–22, 31nn13–14, 33–36, 52n1

naiveté, xi, 33, 36–37, 39, 43–44, 52n7, 109, 143, 150
Nazis, 114, 125n6
Needham, Joseph, 32n22
new world order, 136–137
Ni Lexiong, 56–57, 67, 76n15, 77n16, 88n4
Noddings, Nel, 14, 30n5
non-coercion, ix–x, xiii, 43, 59
non-violence, 44, 53n16, 77n18, 87, 94, 100, 123, 134, 150

O'Neill, Paul, 126n10
Orwell, George, 124

pacifism, xi–xii, 29n3, 38, 41, 55–58, 75n2, 106–8, 150
peace, 38, 61, 83–84, 95, 124, 133, 136–37, 144; through strength, 11, 27–28; through violence, 46, 49, 57, 81, 137; world, 59, 66, 79–80, 134
Powell, Colin, 17, 126n10, 127n17
prohibitions, 17, 25, 46, 48, 59, 109
Project for a New American Century (PNAC), xiii, 124–25, 131–32, 140, 143
propaganda, xii, 30n3, 43, 110, 115, 117–124, 127n16, 129n27, 136
punitive expeditions, xi–xii, 34–35, 41, 49, 56, 58–61, 63–64, 66, 72, 76nn7–9, 83, 85, 88n4, 89n10, 91–92, 94–107, 110, 111nn3–4,9, 112n17

Qin, Cao, 11–12, 29n3, 75n1, 93

remonstrance, xii–xiv, 110, 113, 116–17, 123, 144. *See also* intellectuals, responsibility of

reputation, 20, 36, 89n7,12
revenge, 43, 85, 88n2, 118
rights, 19, 92, 115, 124, 134
Ruskin, John, 14–16, 18, 30n6
Russell, Bertrand, 131, 136, 140, 146nn17–19
Russian bombings, 117–118

security, 75, 85–86, 89n12, 142
self-cultivation, 2, 6, 46, 60, 94–95, 114–16, 129n26
self-defense. *See* war, defensive
selfishness, xii, xiv, 3, 41, 110, 113–116, 123, 125nn1,3, 128nn22,24, 133, 143, 150
September 11, 119–121, 123–24, 127n18, 129nn27–28, 131–32, 145n2
Shue, Henry, 90n15
Simons, Geoff, 66
Slingerland, Edward, 7, 50, 128n20
soft power, 108–9, 150–51
Stalnaker, Aaron, 87n1
Stephan, Maria, 53n16
strength, xiii, 22, 36, 58, 75, 81, 84, 87, 90n13, 104; non-military, 84, 87, 142. *See also* peace, through strength
Stroble, James A., 31n15
Sun Bin, viii, 11, 20–23
Sunstein, Cass, 137, 146n20
Sunzi, viii, 11, 20–24, 30n4, 32n20, 60

torture, 27, 51, 54n21, 93, 120–21, 124, 127nn13–17, 129n28
transcendent principles, x, 18–19, 24–25, 27, 43
true kings, 37–39, 57, 63, 65–66, 83–85, 88n7, 89n10, 91, 96–103, 105–8, 111n5, 112n10, 132, 138
Twiss, Sumner B., xii, 55–57, 73, 76n7, 88n1, 91–94, 97–106, 108, 111nn1–4, 112n10, 112nn15–17, 134
tyranny, xii, 41, 57, 69, 75n1, 76n8, 77n16, 91–94, 97–100, 102–3, 109, 111n1, 136, 143, 145n2

tyrants, 26, 28, 44–45, 57, 59, 62–63, 65, 77n15, 85, 89n10, 97, 103, 143; of ancient China, xix, 39, 41, 62, 67, 76n11, 85, 88n5, 116

United Nations, 66, 76n12, 119, 126nn10–11, 127n17
unopposed, 61, 67, 74, 76n9, 84, 96, 108

Van Els, Paul, 21–22, 31n14, 34
Van Norden, Bryan W., 45–46, 48, 52n1, 62, 76n13, 95–96, 99
Vermeule, Adrian, 137
victory without battles, 20, 23–23, 39, 54n19, 57, 65, 82–85, 97–98, 108–10
violence, cycle of, 37, 43, 85, 88n2
virtù, 12–14, 18
virtuous cycle, 37, 149

Waley, Arthur, 39
Walter, J. H., 15
Wang Liqi, 38
war: antiwar sentiment, 11–12, 16, 19–23, 31n17, 40–43, 80, 149; consequences of, xi, 43–44, 48, 72, 79, 81, 86, 93, 95, 109, 132; defensive, 34, 38, 55, 58–59, 73–75; as duel, x, 12–14, 16–19, 22, 31n15; ex-post facto justification for, 105, 112n17; as masculine activity, 11–17, 19–20, 27, 30n3, 40; to maintain order, 56, 67, 77n15, 80, 88n3; moral limits of, 17–19, 94; *See also* just war criteria; as necessary evil, 31n17, 81; prevention of, 32n18, 34–35, 39, 41, 43, 62, 64, 72, 149; psychological trauma of, 111n2; romanticization of, x, 11–19, 21–22, 27, 149; self-serving motives for, xii, 62, 110, 113, 119, 125n1, 128n19, 132; women, influence on, 14–16; *See also* Gulf wars, and just war criteria
war crimes, 17, 27, 64–65, 81

Watson, Burton, 20
Watts, Alan, 137
weapons of mass destruction, 16–18, 32nn21–23, 105–6, 119, 126n11
welcoming, of military intervention 28, 54n19, 58–61, 64–66, 68, 76n6, 89n7, 97, 104–5, 112n14
Whitehouse, Sheldon, 127n13
Wilkerson, Lawrence, 126n10, 127n17
Wood, Neal, 12
world government, xiii, 131, 133, 136–38, 140, 143, 146n19

Xunzi, x–xiii, 1, 3–4, 6–8, 20–21, 25–26, 28, 36–37, 47–49, 51, 53nn12,14,17, 55–57, 60, 63, 68, 75, 76nn7,10, 79–87, 88–90nn1,3–4,7–14, 91, 93–94, 97–98, 100, 102–10, 112n16, 113–16, 125n3, 128n22, 132–37, 140–43, 145nn5–7,9–10, 146nn14–15, 147nn23,28, 150–51
Xunzian globalism, xiii, 134, 151

Yan Xuetong, 56, 58, 67, 75n2, 134, 140, 144
Yang Qianru, 75n1
Yao, Xinzhong, 26–27, 48, 54n19, 56, 106
Yu, Yi-Ming, xii, 79–82, 88nn2–3

Zhang, Ellen, 23, 40–43, 53n13
Zhuangzi, 1, 20, 53n13, 147n22
Ziporyn, Brook, 31n10
Zuo Commentary, 49

Lightning Source UK Ltd.
Milton Keynes UK
UKHW021409111022
410300UK00002B/25